Managing Business in the Civil Construction Sector Through Information Communication Technologies

Bithal Das Mundhra
Simplex Infrastructures Ltd, India

Rajesh Bose
Simplex Infrastructures Ltd, India

A volume in the Advances in Business Information Systems and Analytics (ABISA) Book Series

IGI Global
PUBLISHER of TIMELY KNOWLEDGE

Published in the United States of America by
IGI Global
Business Science Reference (an imprint of IGI Global)
701 E. Chocolate Avenue
Hershey PA, USA 17033
Tel: 717-533-8845
Fax: 717-533-8661
E-mail: cust@igi-global.com
Web site: http://www.igi-global.com

Library of Congress Cataloging-in-Publication Data

Names: Mundhra, Bithal Das, 1942- author. | Bose, Rajesh, 1981- author.
Title: Managing business in the civil construction sector through information communication technologies / by Bithal Das Mundhra and Rajesh Bose.
Description: Hershey : Business Science Reference, 2020. | Includes bibliographical references and index. | Summary: "This book highlights the role and use of computer-aided design and tools in the production activities of civil construction companies"-- Provided by publisher.
Identifiers: LCCN 2020001207 (print) | LCCN 2020001208 (ebook) | ISBN 9781799852919 (hardcover) | ISBN 9781799852926 (paperback) | ISBN 9781799852933 (ebook)
Subjects: LCSH: Construction industry--Technological innovations. | Construction industry--Management. | Building--Data processing.
Classification: LCC HD9715.A2 M86 2020 (print) | LCC HD9715.A2 (ebook) | DDC 624.0285--dc23
LC record available at https://lccn.loc.gov/2020001207
LC ebook record available at https://lccn.loc.gov/2020001208

This book is published in the IGI Global book series Advances in Business Information Systems and Analytics (ABISA) (ISSN: 2327-3275; eISSN: 2327-3283)

British Cataloguing in Publication Data
A Cataloguing in Publication record for this book is available from the British Library.

Advances in Business Information Systems and Analytics (ABISA) Book Series

ISSN:2327-3275
EISSN:2327-3283

MISSION

The successful development and management of information systems and business analytics is crucial to the success of an organization. New technological developments and methods for data analysis have allowed organizations to not only improve their processes and allow for greater productivity, but have also provided businesses with a venue through which to cut costs, plan for the future, and maintain competitive advantage in the information age.

The **Advances in Business Information Systems and Analytics (ABISA) Book Series** aims to present diverse and timely research in the development, deployment, and management of business information systems and business analytics for continued organizational development and improved business value.

COVERAGE

- Geo-BIS
- Business Systems Engineering
- Business Intelligence
- Forecasting
- Information Logistics
- Strategic Information Systems
- Statistics
- Management Information Systems
- Data Management
- Business Process Management

The Advances in Business Information Systems and Analytics (ABISA) Book Series (ISSN 2327-3275) is published by IGI Global, 701 E. Chocolate Avenue, Hershey, PA 17033-1240, USA, www.igi-global.com. This series is composed of titles available for purchase individually; each title is edited to be contextually exclusive from any other title within the series. For pricing and ordering information please visit http://www.igi-global.com/book-series/advances-business-information-systems-analytics/37155. Postmaster: Send all address changes to above address.

Titles in this Series

For a list of additional titles in this series, please visit:
http://www.igi-global.com/book-series/advances-business-information-systems-analytics/37155

Handbook of Research on User Experience in Web 2.0 Technologies and Its Impact on Universities and Businesses
Jean-Éric Pelet (ESCE International Business School, INSEEC U Research Center, Paris, France)
Business Science Reference • © 2021 • 300pp • H/C (ISBN: 9781799837565) • US $295.00

Empowering Businesses With Collaborative Enterprise Architecture Frameworks
Tiko Iyamu (Cape Peninsula University of Technology, South Africa)
Business Science Reference • © 2021 • 309pp • H/C (ISBN: 9781522582298) • US $195.00

Natural Language Processing for Global and Local Business
Fatih Pinarbasi (Istanbul Medipol University, Turkey) and M. Nurdan Taskiran (Istanbul Medipol University, Turkey)
Business Science Reference • © 2021 • 452pp • H/C (ISBN: 9781799842408) • US $225.00

Applications of Big Data and Business Analytics in Management
Sneha Kumari (Vaikunth Mehta National Institute of Cooperative Management, India) K. K. Tripathy (Vaikunth Mehta National Institute of Cooperative Management, India) and Vidya Kumbhar (Symbiosis International University (Deemed), India)
Business Science Reference • © 2020 • 300pp • H/C (ISBN: 9781799832614) • US $225.00

Handbook of Research on Integrating Industry 4.0 in Business and Manufacturing
Isak Karabegović (Academy of Sciences and Arts of Bosnia and Herzegovina, Bosnia and Herzegovina) Ahmed Kovačević (City, University London, UK) Lejla Banjanović-Mehmedović (University of Tuzla, Bosnia and Herzegovina) and Predrag Dašić (High Technical Mechanical School of Professional Studies in Trstenik, Serbia)
Business Science Reference • © 2020 • 661pp • H/C (ISBN: 9781799827252) • US $265.00

Internet of Things (IoT) Applications for Enterprise Productivity
Erdinç Koç (Bingol University, Turkey)

For an entire list of titles in this series, please visit:
http://www.igi-global.com/book-series/advances-business-information-systems-analytics/37155

IGI Global
PUBLISHER of TIMELY KNOWLEDGE

701 East Chocolate Avenue, Hershey, PA 17033, USA
Tel: 717-533-8845 x100 • Fax: 717-533-8661
E-Mail: cust@igi-global.com • www.igi-global.com

Table of Contents

Preface

Construction industry has a pivotal role to play in building a robust infrastructure of India. That the figures of GDP of our country are in proportion with those of the construction industry, is in evidence in authoritative reports published by government bodies and concerned agencies. It is in this context and in the face of increasing competition within the industry itself, that a construction company needs to stay focused with a critical eye on emerging threats and opportunities that a developing economy accords. Long before this book was written, stalwarts and leaders of the construction industry had realized that Information Technology has a vital role to play in the evolution and strengthening the position of construction companies and the industry as a whole. Does this, therefore, make this book any less relevant? On the contrary, there is a need to dive deeper into the innards of civil construction by holding hands with Information Technology. The ways that Information Technology is important and indispensable are manifold within this realm. Computer-aided design and visualization; building information modelling; computer aided cost estimation; construction engineering and management; road design and transportation management; hydraulic engineering; structural engineering; geo-technical engineering; computer-aided facilities management; business information management; and electronic mail are the routes through which Information Technology can and does boost the operational efficiency of construction industry. The composition of this book will allow readers to gather insights into evolving computing technologies such as the Internet-of-Things, cloud-based computing, next-generation of cloud-based ERP solution, artificial intelligence, and blockchain in the context of industrial sectors, including the construction industry, in India.

One of the key objectives of the authors while preparing the initial drafts of the book, was to accord readers a window into the world of civil construction industry in India as it appears through the prism of an information technology specialist. The book highlights the use of computer-aided design and tools, and the vital role that such forms of information technology has to play in the actual production activities of any civil construction company – be it large or small. Subsequent chapters focus on equally vital areas such as that of construction management, contracts management, materials management, human resource management, and enterprise resource planning. Chapters on cloud computing technology, Internet-of-Things, and artificial intelligence will enable readers to acquire overview and grasp the basics of these latest trending technologies as of writing this book. This book has been written for students, entrepreneurs and professionals by authors with considerable years of experience in implementing modern information technology and tools in a large prestigious construction company.

The core focus of the authors is to provide a sound foundation on which readers can expect to apply their knowledge in this field. The book was written primarily with a two-fold objective. The first one being to enable students of civil engineering have an overview of the various information technology tools that are employed in a civil construction enterprise. More importantly, students shall be able to understand the key areas where IT helps civil construction companies to function in a way that helps in staying competitive. The second objective of the book was to aid entrepreneurs and senior executives with direct stake in civil construction engineering. The book attempts to provide an overview on the various IT tools and allied technologies that are essential for the functioning of a civil construction company in a competitive market. The book includes chapters on cloud computing technologies and Internet-of-Things (IoT). These have been written to support businesses and senior executives when initiating major policy decisions, and also help them understand how such state-of-the-art advancements can favourably affect transitioning from traditional IT systems to more agile system of data and information management on cloud.

IT consultants looking forward to venturing into the world of civil construction engineering will find that this book presents a broad perspective on the various aspects of civil construction activities in which information technology is relevant. Consultants acting as independent observers or hired by businesses to implement appropriate strategies to modernize their operations through infusion of IT tools and technologies, should be able to formulate their own course of action based on a reading of this book.

In the course of working on the book, the authors realized that an element of fluidity ought to be injected in the book so as to make it a thoroughly interesting read for students, professionals, consultants and business owners alike. As authors with more than a passing interest in this domain, the fact that the latest trends in information technology has much to offer in the field of construction engineering, is an aspect that was worth writing on. While the book leaves scope for improvement in its present avatar, it was never intended to be a wholesome text book. On the contrary, the book was written bearing in mind the quick reference needs that a student, a business owner, or even a consultant might have in mind to set one's course in a definite area where IT tools make an impact in the civil construction sector.

The book is divided into 10 distinct chapters. While we have tried to maintain a degree of insularity in each of the chapters for the sake of modular reading, that has not always been possible. And with good reason. Our own experiences in introducing new IT tools and technologies in our enterprise have constantly thrown us challenges to build a synergy between IT systems. To stay profitable, the different arms of an enterprise must cohere and work as one. So, too, must the different IT tools that are used by each department, section, or branch. Engineering students will be introduced to the current production and management areas where information technology tools and systems are used in large- and medium- construction companies. That is not to say that smaller construction companies do not use information technology tools in the key areas of design drawing, project management, ERP, or human resources. Medium- and large-scale civil construction companies rely on information technology in almost all areas of their operations beginning from production, planning, accounting, stores and inventory management. This book provides a bird's eye view of the major areas where IT is applied without going into the mechanics of implementation and use of such tools and technologies. For senior executives, business leaders and entrepreneurs in the civil construction sector, the book offers a holistic view of how information technology can help in making key strategic decisions for sustained business operations. For them, this book can assist in identifying the key points on which to base their further research and studies.

Acknowledgment

The authors would like to gratefully acknowledge the contributions of all at Simplex Infrastructures Limited, a company renowned for successfully completing prestigious civil and heavy engineering construction projects in India and abroad. Many of the chapters could not have been written without feedback from civil engineers, planners, designers, HR professionals and from members of MIS, ERP Functional, Oracle Developer and Data Center IT support teams of the company.

We are indebted to Mr. Apurba Mukherjee who helped germinate the idea of writing a book on contributions of Information Communication Technology in Indian Construction Sector. We are especially thankful to Ms. Jaya Bhattacharyya, Sr. Vice President and to Mr. Tapash Sengupta, Principal Consultant (Systems Integration), for their invaluable assistance in ensuring that we did not lack necessary equipment or resources while working on this book project. It was under their leadership and guidance that the company's Data Center was setup more than a decade ago, and has since been running successfully supporting a unique ERP system developed in-house for the company. We would also like to mention the work of Mr. Satadru Sengupta, IT Department, who put together text and data from research materials made available to us. Special thanks for Mrs Swati Bose who assisted in formatting and designing.

We would also like to thank all other departmental heads of the company who supported us and provided necessary materials to base our research on. Without their invaluable inputs and assistance, many of the chapters in the book may have remained incomplete.

Section 1

Impact of Information Communication Technology in the Construction Industry

Chapter 1
Information Technology and Construction Industry

ABSTRACT

For thousands of years, construction engineering has played a vital role in the advancement of human civilization. Humans have always aspired to build taller, grander, and more robust constructions either for civilian purpose or to defend cities and towns from invading armies. The architectural splendour of the pyramids in Egypt and the Great Wall of China are not only awe-inspiring examples for the engineering feats of the ancients but also remarkable for the robustness in evidence for having stood the test of time for over thousands of years since these were built. Modern civil construction engineering, too, requires optimum resource utilization, speed of execution, and adherence to best engineering practices. The rapid strides made in information technology makes it possible for architects and engineers to achieve those goals. This chapter discusses the role and impact that information technology has in the domain of civil engineering and construction sector.

The influence of information technology in construction industry today is undeniable. Software technology and tools have made it possible to process data from multitude sources and present such into meaningful information for humans. The construction industry, like any other, has reaped the benefits of utilizing information technology. Construction companies need to operate remote sites producing disparate sets of data. Computers and information technology makes it possible to perform complex tasks to process extremely

DOI: 10.4018/978-1-7998-5291-9.ch001

large volumes of data flowing through a typical organization in construction sector. Today, construction companies use a wide variety of software tools that address requirements of architectural designs, civil engineering drawings, project planning, financial management, procurement, quantity surveys, site management, contracts, valuation, marketing, etc. The world of information technology has completely revolutionized commerce and industries across all sectors are conducted. The construction sector has been no exception to the influence that information technology has cast on trade and manufacturing across the globe.

The initial acceptance of information technology in the construction scene was sedate if not a complete crawl in comparison to comparable industries in infrastructure and manufacturing sectors. For most part in the 1980s, personal computers were used in handful construction companies. Construction firms mostly used personal computers for administrative functions related to accounts, wages, salaries, etc. Very few companies, back then, used information technology to plan their production schedules or digitize their core operations. By late 1980s, a growing number of construction companies began using personal computers to produce digitized versions of engineering drawings, use spreadsheet and word processing software, as well as use computers for planning and monitoring. The growing reliability and personal computers made it possible for construction companies to begin experimenting with the idea of allowing information technology to slowly aid in complex data processing functions. It was not long before businesses and organizations engaged in civil engineering and construction activities realized the need to adopt information technology. The highly complex communication structure in civil engineering projects, and the intricate details which are involved in project management of a construction project, cried out for a systematic and robust system to manage and monitor progress and data flow. While information technology has now gone on to considerably ease the burden of managing vast array of construction processes involved, there have been gaps in adoption of information technology itself by civil engineering and construction companies. Productivity gains from information technology had not been clearly in evidence initially. As modern information technology evolved and allowed integration between software applications for data and information sharing, companies engaged in construction activities gradually began to see gains in productivity. It is without doubt that there remains a scope for improvement at any level. In a competitive world, construction companies can ill afford to ignore emerging trends in information technology. Computers and software allow for better integration of production processes and, thus, help directly improve efficiency. Experts and researchers have observed that productivity levels in the India construction sector have by and large stagnated over a period spanning at least 40 years. By contrast, construction sectors in other countries that have been quick to adopt information technology, have witnessed surge in productivity levels. The apparent slow pace in adopting information technology by Indian construction companies can be attributed to the fundamental nature of resisting change by the Indian construction industry as a whole. The level of adoption of information technology slowly began to gather pace as more and more engineering students began to be exposed to higher education that advocated and promoted the use of latest in computer technology and associated tools.

Thus, it is not without reason that higher education is considered to be prime driver in the constant quest to improve production processes of construction companies. It is, therefore, of the utmost importance that students of civil engineering courses and degree programmes be introduced to the latest and emerging trends in information technology. As such, there is no room for doubt that if students are trained and methodically introduced to latest in computer hardware and software advancements, they would be the ones to lead the charge in transforming the way construction companies operate, and build a competitive edge hitherto unrealized. This book explores the need, the potential impact, and the implications that information technology holds for the construction sector in general and for the Indian construction industry in particular. It examines closely and takes the reader on a journey through the various facets and features that modern information technology has to offer, and those which can be woven into the very fabric of a construction sector enterprise.

DEFINITIONS OF "CONSTRUCTION" AND "INFORMATION TECHNOLOGY"

The term "construction" generally describes activities that are related to erecting buildings, plants, roads, bridges, etc.; or the very structure of a building itself. Unlike in most other manufacturing processes, civil engineering production usually takes place at the very site itself where a structure is to be located and used. Projects involving erection and commissioning of such civil engineering artefacts can stretch over extended lengths of time. By contrast, most other manufacturing processes in other engineering sectors require fraction of the time involved. When we speak of civil engineering construction process, we think in terms of creation of designs and drawings, and the final construction process itself. However, for a comprehensive definition of construction to take root in our minds, it is necessary to discuss the operations and maintenance aspects. Information produced during design and construction phase, are extremely vital for the operations and maintenance of a building, a plant, a road, or a bridge throughout the erection stage and beyond. Traditional construction process models often do not include information on the manufacturing of the very building materials required for construction. Rarer still are traditional construction process models known to include planning and inspection activities. Despite this, it still is important to include these factors to produce a holistic construction process to move decrease overall maintenance costs and time involved to effect maintenance and repairs.

The term "information technology" is deemed to mean all forms of computers; associated equipment and peripherals; software; storage systems; firmware; transfer and process of data and information; technologies and services that support or otherwise govern the manner in which computers, peripherals, software, and services are conceptualized, manufactured, and deployed. In the initial years that saw personal computers slowing gaining popularity, the term electronic data processing, or EDP, was quite common. That term was closely associated with data processing using computers for the purpose of storage and presentation of data in assorted formats. This term is slowly falling into disuse as information technology has progress far beyond those days when data processing was prevalent throughout organizations irrespective of the sector or size. Computers of today are no longer to be found within confines of air-conditioned rooms that were also often sterile and kept dust-free as far as possible. Information Technology has today progressed to a stage where physical dimensions of computers have shrunk, and are a far cry from the days when computers were only to be thought of as huge machines housed in large air-conditioned halls. The smartphones or tablet PCs now has more processing power per device than the predecessors of the modern personal computer. The average person carries enough computing power and software applications in one's pocket than what would have been considered possible even a decade ago. We have grown ever so reliant on information technology in all facets of our daily lives that to think of spending a day without our mobile devices, which are more than double up as true personal computers, can throw our routine out of gear.

The definition of Information Technology, thus, does not exclude our phones. Nor does it make a distinction insofar as photocopying machines, telefaxes, or scanners are concerned. These devices that "talk" to one another across wires and networks are as much a part of information technology as traditional computers and allied systems are deemed to be. The modern avatars of phones, faxes, etc. are produced by their respective manufacturers in a way that these can be seamlessly connected to computers and each other through wired or wireless means. Software applications and firmware integration now make it possible for most laptops, personal computers, and phones to send and receive faxes as well as emails. Thus, it does not come as a surprise to observe the gradual and often inextricable link up of information technology with construction.

THE DEVELOPMENT OF INFORMATION AND COMMUNICATION TECHNOLOGY IN CONSTRUCTION

The gradual proliferation of information technology and tools in the constructor sector has had a positive effect. Over time, the construction sector has generated new employment opportunities and jobs that involve managing assorted activities that include, but are not limited to, collaboration, coordination and exchange of data between independent entities that jointly work on a construction project. Few technical systems, that are complex in nature, evolve full formed. Rather, such technical systems initially rise and take shape from a localized state. This, at times, is noticed to have certain unintended consequences – chief among which is that the evolution of such technical systems is often found to be unbalanced in terms of automation. Such state leads to undesirable consequences and unintended silos of automation. As a result, automation (McCullouch, 1997) is perforce made to link up with products of manual processes; thus, neither being productive nor cost-effective. This phenomenon is known to occur in the case of computerized information systems, too. The evolution and modernization of information technology has led to development of software that are used almost universally for data analysis and manipulation. Many of such software tools, though eminently capable of processing immensely complex data sets, were often developed in isolation. Although, these possessed phenomenal number-crunching features (Cheok, Lipman, Witzgall, Bernal, and Stone, 2000), that were developed in 1960-1970, ranging from finite element to critical path analyses, the flow of information was still far from being a completely digitized one. Notwithstanding the advanced features for the time, information technology that was used during the early days by construction companies, involved producing engineering drawings (Liu, 1997). that had to be first plotted on paper to enable information exchange between disparate application software.

Considered as an important milestone in technological evolution after the first printed book, the development of scaled technical drawings produced through software has allowed free and unimpeded exchange of engineering data and information between different software applications. It was during the 1970s that design and development of computer graphics technologies began to gather pace. The early versions of computer-aided designed (CAD) software were developed (Thorpe, and Mead, 2001) using novel computer graphics technologies. However, despite their inherent sophistication for the time, these CAD software applications operated independently by relying on traditional methods to produce and exchange data. Interestingly, large public sector players during that period began exploring possibilities to develop information systems (COMIT: Construction Opportunities for Mobile IT, 2005) that would allow them to adopt standardization and prefabrication. Soon, software developers and information technology companies began to show interest in developing computer-aided software systems to meet those requirements. However, it was not long before the momentum was lost as the demand for such systems shrunk as those large public sector construction programmes (Prasad, Ruggieri, 2003) were gradually withdrawn. With the advent of 1980s, a new wave in computer technology revolutionized the way data and information were managed, processed and stored. The personal computer (Wikipedia, 2006) offered far more processing power at fraction of the cost it took for the previous generations of computers. As costs plummeted, personal computers began to be widely accepted and adopted by organizations for their offices, and by individuals for personal use. But more significantly, for the first time ever, remote project sites of construction companies could be equipped with serious equipment worthy enough to be deemed as a force-multiplier. Complex and larger sets of data could be now processed quickly and efficiently by remote project site offices using personal computers. While information technology began to morph and mould, communication technologies, too, saw notable progress and advancements being made. Notably, the telephone evolved into mobile phones that are, today, far removed from the telephones that were tethered to wires and were to be found in abundance even a decade ago. The development of local and wide area networks, coupled with wireless transmission technologies, entirely transformed the way communication technologies and information technology was linked. In fact, the popularity of the internet starting from 1990, matched by surging demands for smartphones, ushered in a new paradigm in how data flow and information began to be processed in the last few years. The effects of the radical change evident today in the form of complex data processing and presentation of information using smartphones, internet, and ever-evolving information and communication technologies have thrown up a host of challenges. The construction industry, with their myriad and complex requirement that govern flow of data and information, has been influenced and affected as a result as any other manufacturing industry. On the other hand, this has opened up newer opportunities as demand for professionals and engineers trained to handle sophisticated software applications and IT tools in civil construction sector, has grown.

WHY NEED IT IN CONSTRUCTION

The construction industry functions on tight deadlines. For a project to be on the right side of financial operations one of the key requirements is that flow of data and information should be unhindered and accessible to relevant stakeholders on demand without delays. Since there are several activities and processes that are part of any construction project, the dynamic nature thus exhibited makes it imperative that flow of information is always treated on priority basis. Delays in transmission of information can and often do adversely affect production schedules and cause cost overruns. Given that projects are now being executed on reduced profit margins to stay competitive in the market, paper-based documentation is often observed to be woefully inadequate in coping with the rigorous demands for complete and on-time flow of information.

The major shortcoming of paper-based documentation is the amount of time involved transmitting data or information from one site to another, to central office or vice versa. As a result, there is a general lack of cohesion, and often there is confusion with the relevancy of data received at any given point of time. This happens because data transmitted in such fashion may have been superseded and modified by the time the transmitted version reaches the recipient. It has been observed that field personnel and supervisors at construction project sites may spend up to 50% of their working hours collecting and analysing data. Around 2% may be actually spent on construction sites for manually tracking and collating data on progress of work.

As operational efficiency depends on the accuracy of data collected, the veracity of the data collected is influenced by report-preparation skills and judgement of the people actually engaged in collecting and collating the data under actual field conditions. Since, in most cases the data collected are not captured digitally, i.e., through use of smart devices supported by appropriate computing systems, the process of data transfer itself from a remote location is often time-consuming and requires considerable man-hours to be dedicated for the purpose. Owing to manual processes involved, the chances of inadvertent errors creeping in without being apparent or obvious, are quite high. The matter is further aggravated with propagation of the errors throughout the organization as the data and information branches out and cascades. To empower project teams with accurate and timely information, the only viable way forward for construction companies is through adoption of state of art information technology.

THE INTRODUCTION OF IT IN CONSTRUCTION FROM A HISTORICAL PERSPECTIVE

From a historical perspective, information technology, in its infancy, was viewed as a support service necessary for data processing. Analytical programs that ran structural analysis calculations were dependent on repetitively tiresome manual process of preparation of input data using, in most cases, punched cards. Although, in the years leading from 1980, witnessed the gradual acceptance and use of computer-aided design (CAD) software, the emphasis lay squarely on creation and viewing of certain select set of data and information. Initially, a CAD software would be first installed on dedicated minicomputer. Viewers would mill around a single screen to observe an image during design-time. Further down the line, information retrieval would be made possible using several terminals connected to the minicomputer that had the CAD software installed and running. Information on the designs created could subsequently be obtained in print form using expensive A1/A0 plotters. The plotted diagram, if required – which was more often than not, then sent for making appropriate number of copies before distribution to concerned departments, consultants, drafts people, and engineers. Quite naturally, such an arrangement failed to be a hit among users or by organizations. Despite being favourably advertised by software developers and vendors, CAD software, then, never could manage to find a firm footing or mass following among consumers.

In actual practice, designers can spare barely a couple of hours per day creating complex drawings. The bulk of their hours are used up with retrieval of information, communicating with co-workers, pause for creative reflection, and carry out any other activities directly unrelated to actual production of engineering drawings.

Today, the situation has undergone a sea of change as advancements made in network technology, the Internet, video-conferencing, and smartphones have significantly facilitated in retrieving information and engaging in communication efficiently and accurately.

THE LEVEL OF INFORMATION TECHNOLOGY USAGE IN THE CONSTRUCTION INDUSTRY

To assess the overall effectiveness of information technology at work in construction industry, it is often found useful to begin with determination on the extent use of information technology at various levels. Looking at measuring the effectiveness of information technology from this perspective, there are broadly three levels. These levels, classified from bottom up, are designated as operational level, tactical level, and strategic level [Figure 1]. At the base level, all operational activities of a firm are undertaken by all associated users in construction enterprise. The role of information technology is to provide support to users in order to better engage in business operations. At the tactical level, where business operations are managed by mid-level manager, information technology and tools help in making quicker and more effective decisions. As would be evident from the pyramidal view of the figure shown, information technology can help deliver smooth flow of information across multiple levels systematically.

Figure 1. The levels of IT usage in a construction industry

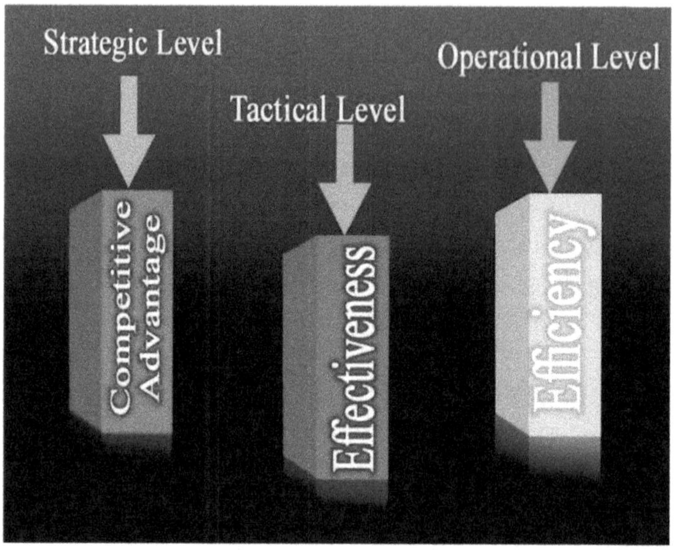

At the top of this pyramidal view, sits the strategic level. This level represents senior management of an organization. For the senior management, information technology functions an extremely vital role by collating information from across organization. By presenting information in easy-to-read and readily interpretable format, information technology enables to make effective strategic decisions that can push an organization up the ladder ahead of its competing peers in the market. In the case of construction companies, the stage at which strategic decisions are made, have turned ever so more significant as global and national markets grow stiffer with the addition of newer entrants. Information Technology can help construction companies formulate effective strategies and ensure a cycle of continuous integration, deployment, and feedback.

IMPACTS OF INFORMATION TECHNOLOGY USAGE IN THE CONSTRUCTION INDUSTRY

One of the more difficult challenges that construction companies face in implementation and use of information technology and tools, is uniformity in application. To be able to analyse the impact and effects of information technology, it is necessary at first to define the effects that information technology has on industry. To illustrate this approach, a layout showing varied effects of information technology in an organization is presented in Figure 2.

The effects can be grouped into three groups of technical, economic, and strategic as seen in Figure 2. Impacts clubbed under technical aspects are considered to be positive effects for organization. These effects are borne out of the use of information technology at the operational level involving prompt corrections, posting of current information, proper storage of data, appropriate information sharing, easing of communication, bringing about reduction in costs, and progressively lowering consumption of paper. As such, these changes at the operational level stimulate increased revenues, spur improvements in quality, and expand scope for value addition; while lowering resource consumption. On the flip side, the associated economic impact of information technology is described as being negative. Expressed in terms of continual and periodic upgrades to IT infrastructure, the costs of investment and training imparted to users are considerable.

It must, however, be borne in mind that the positive influence brought about by information technology does not necessarily translate to competitive advantage for an organization. For an organization to meet market challenges head-on, both technical and economic effects need to be fused to bring about a strategic impact. When information technology (De la Garza, Howitt, 1998) is used by senior management for setting up of definite goals, charting a course for formulating and implementing of strategies, strategic impact is created. If strategic goals are set using information technology, an organization can be in a position to enhance its competitiveness and aspire to a more advantageous position than its competitors. For construction companies that project themselves as a global player and bid for contracts in the international market, attention needs to focus on Information Technology in their quest to build up effective strategies.

Figure 2. The different impacts of IT on the firm

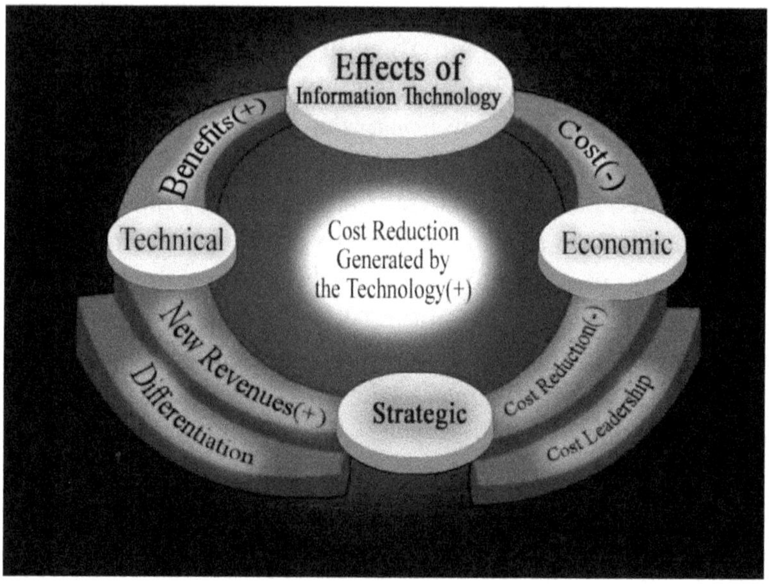

APPLICATION OF INFORMATION TECHNOLOGY IN THE CONSTRUCTION INDUSTRY IN INDIA

The influence of Information Technology in the construction sector has progressed rapidly. No longer is IT confined to office productivity suites, computer-aided designing software and tools, cost evaluation software, software for quantity calculations and quota management. With rapid proliferation of the Internet, modern Information Technology and tools have changed the landscape of operations within civil construction companies. Office productivity tools, that are the most popular form of software among consumers worldwide, have grown in popularity and sophistication. Aside from being used for usual office and contract administration processes, modern versions of such software enable building documents to be exchanged in electronic form.

Already in its most common place implementation involving office productivity software, Information Technology has established itself capable of improving productivity and facilitating in keeping costs under control. Among the profusion of computer software types that is used in civil construction and engineering, computer-aided design software stands head and shoulder over most. Engineers, architects, consultants and designers working in civil construction engineering use some form of computer-aided design (CAD) software from a select corps of software manufacturers. Software tools that help in cost evaluations of projects are also found to be in demand. Entrepreneurs and project planners rely on Information Technology to prepare and present competitive rates during bidding processes. In such cases, quota management software plays a significant role in reducing costs while presenting an overall view of quotas and bills of quantities. The disadvantage of quantity calculation software is the limitation in identifying drawings. Ideally, once components and respective sizes are noted, such software should be able to automatically calculate quantities and produce relevant bills of quantities. Scanned drawings can be converted to raster images for software to determine components and corresponding sizes. Similarly, the volume of steel required can be computed based on drawings scanned. By far, however, the area where Information Technology has registered the quickest growth in construction sector has been in establishment of internal and network communications for data and information exchange. Without network communications, teams would have found it increasingly difficult to work on complex building projects. In practice, it has been observed that the presence of robust and reliable network communications fosters an environment of teamwork and coordination.

As an example, central offices would find it easier to compare and monitor actual figures transmitted from production sites with projected figures. It also becomes far easier to access rates of input materials instantly to enable immediate procurement processes and, thus, take advantage of competitive pricing whenever possible. Network communications have also facilitated construction companies to share data on previously executed projects with internal departments as well as with external agencies, and cooperating organizations and vendors.

THE ROLE OF INFORMATION TECHNOLOGY IN CONSTRUCTION INDUSTRY

It has been opined that success of inducting modern Information and Technology tools in construction sector has paved the way for progressively more investments being made in this sector for purchase and installation of modern computer hardware and software. The invention of personal computers in 1980s, ushered in a new wave of technological evolution that thoroughly revolutionized the way businesses and even lay people on the streets communicated and transacted. Today, Information Technology is an inalienable feature that features prominently in significant investments that businesses need to make to stay competitive, and to perform productive, planning, and administrative activities. The growth of Information Technology in construction sector is expected to rise as many times over as in any other manufacturing sector. Businesses have not lagged behind in realizing that Information Technology is an asset that facilitates in staying focussed and relevant in a highly competitive market. As such the demand and application for newer IT tools and allied technologies are constantly in place with construction companies. It is often argued that the extent of application and introduction of Information Technology in a construction organization is a measure of effectiveness. Smaller players in the construction business can gain traction by introducing cutting edge IT tools and allied technology after conducting appropriate studies. Globally, the construction sector has been witnessing increasing reliance on Information Technology.

In the past three decades, the scope and applicability of Information Technology has evolved within industries. Construction companies rely extensively on Information Technology to simulate and analyse the performance of their designs. In the initial years, the role of Information Technology was confined to accounting, payroll, and for performing analysis on complex calculations. The second generation of computer applications saw decision-based supporting systems coming to the fore. The current and third generation of applications now focus on processing data related to strategic information systems that are considered vital for the very existence of a construction company in highly competitive market. Arguably, the allied revolution that has washed through communication technology has also proven to be extremely significant for organizations. While Information Technology is a force-multiplier and is seen to aid in productivity, the copious quantities of data and information produced are required to be housed within database systems and storage repositories for knowledge sharing. With information being processed and stored in ever increasing troves, businesses have had to pause and rethink traditional forms of IT applications and related hardware. Since, production processes are predicated upon a thorough and effective approach to managing resources such as plants, equipment, machinery, finance, technology, transportation and information, construction companies have had to craft an entirely different approach. To achieve both short term goals and longer term objectives, construction companies needed to focus on operational aspects narrowed down to a micro level of functioning. For example, data fed into programme schedules by site managers are required to be collated

Figure 3. Illustration of how ICT can support the construction process

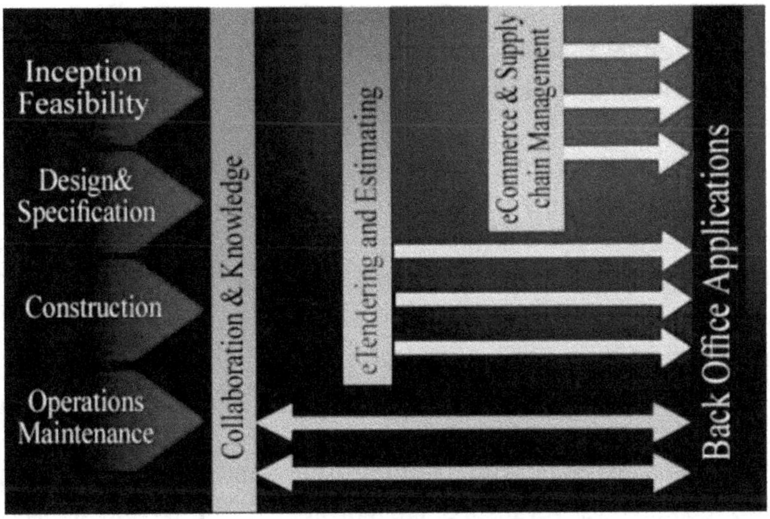

with master schedules and centralized project plans. Suppliers, supervisors, contractors, and other stakeholders simultaneously are also required to update their information and schedule, accordingly. Time sheets, target projections and acquisitions all merge into broader canvas of operational projection and planning. Quality standards can be closely monitored and set in terms of both human resources, materials, and constant updates to project planning. It was a commonly held belief, and in certain cases it still remains to date, that Information Technology is ill-suited to construction and project sites due to the physical nature of work involved. This could not be more further from the truth as resource planning, utilization and management are best served through infusion and appropriate use of Information Technology.

Information Technology is used in a range of tasks by civil construction firms. These can broadly range from producing and managing technical drawings, project planning and monitoring, accounting and reporting, to data and information sharing. However, it is important to note that Information Technology has to permeate to the level of site supervisors and foremen to enable businesses to view real-time data. Only with the help of proper and adequate IT tools would businesses be able to reduce information gap that is often found to exist between sites and offices. In construction circles, Information Technology has the capability of influencing the efficacy and capabilities of construction processes that take place.

Information Technology already helps support a gamut of operational functions at a project site. Software has been developed usually as a response to particular challenges that organizations come across. The broad classification of software is listed as follows:

1. Computer-aided design
2. Building engineering applications
3. Computer-aided estimation
4. Planning and Management of sites
5. Monitoring construction projects
6. Business and Information management
7. Computer-aided Facilities Management

Figure 4 demonstrates how available information technology and tools can be applied across several stages in a project. The diagram clearly outlines the areas where software packages are often found operating independent of each other and discretely within a construction supply chain. The associated table lists the software used in the processes illustrated in the diagram.

Table 1. Essential Software

Vendor	Software	Application	Native Data Format
Autocdesk	AutoCAD	CAD	Dwg (DraWing)
Graphisoft	ArchiCAD	CAD	pln
Autocdesk	Revit	CAD(Parametric)	Rvt(ReViT)
Bentley	MicroStation	CAD	Dgn(DesiGN)
Microsoft	Excel	Estimeting (Spreadsheet)	xlcx, xlx
Sage	Timberline	Estimeting	Pee (PrEcision Estimation)
Microsoft	Project	Planning	Mpp(Microsoft Estimation)
Oracle	Primavera	Planning	Mpp(Microsoft Projrct Plan)

Figure 4. Project model for construction

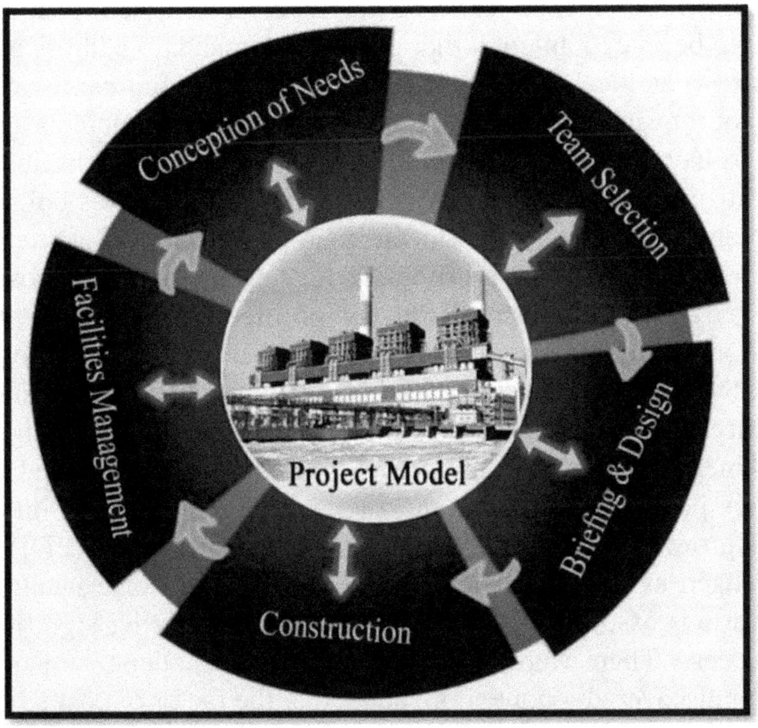

Business and Information Management

The construction industry is one of among high-information churning ones in the world today. Tremendous volumes of information and data are produced. The primary reason for this is that a construction company usually involves a number of suppliers, consultants, associated agencies, its own departments, and of course the offices of its clients. Such disparate and assorted masses of data and information can only be managed in a systematic manner through use of Electronic Document Management (EDM) software. Construction companies of all sizes need to periodically review their business processes and take stock of new advancements made in information technology on a regular basis. Real world case scenarios have adequately demonstrated the efficacy of Enterprise Resource Planning (ERP) software in synergizing various arms and departments that exist within a civil construction company. Another important paradigm in enhancing overall operations and level of training among personnel in a civil construction company is Knowledge

Management (KM). This is a system whereby knowledge acquired either through expertise or experience of employees concerned is documented and filed to be shared among other users, on demand, to address problems. The system is an ideal facilitator in such cases and significantly lowers the quantum of time and resources required for addressing problems and issues that would have otherwise may lead to costly escalations both in terms of time and material. Success of construction projects involves collaboration among teams with specific tasks within an organization. Such teams may either be at site, or situated in offices, or may even be spread across remote corners geographically apart. With the proliferation of mobile Internet, Information Technology has been able to immensely multiply productivity of users. Software exist in the market that cater to such demanding needs. Prominent examples include 4Projects, Sarcophagus, etc. The once ubiquitous fax machines, modems or even internet faxes, have been replaced by email and hugely popular smartphone messaging apps. As data and information began to grow exponentially, Electronic Data Interchange (EDI) became acknowledged as reliable, quick and systematic form of exchanging data among various associated stakeholders without incurring significant loss in the process. There were, however, significant challenges involved that were soon discovered – none more acute that the factor of cost to carry out data interchange. At the turn of the century, the rise of eXtensible Mark-up Language (XML) displayed remarkable promise of achieving a systematic form of data exchange mechanism. Suppliers and buyers could now easily trade complex data and information quickly and conveniently. Along with ERP software marketed by SAP and Oracle, XML technology ranks as one of the most significant achievement in Information Technology that has left an almost indelible impression on almost any manufacturing industry as complex as civil construction engineering.

Computer Aided Design and Visualisation

Before Autodesk introduced AutoCAD software, designs were drawn on paper. Computer-aided Design (CAD) software radically changed the way designs and drawings were produced, viewed, edited, and shared. Several CAD software exist in the market, and all with a common goal of extending users the benefit of producing designs and drawings as cleanly and as efficiently as possible. CAD software have since progressed from 2D to 3D modelling. Designers can now produce digital versions of entire building drawings and test each component and structure in 3D. The plain interface and drawings have now been replaced with rich and realistic images that can be animated to produce effects that mimic closely real-world situations. Building Information Modelling (BIM) is the new mantra in civil construction engineering. With BIM, construction companies can greatly enhance the collaboration among different stakeholders to enhance productivity.

Figure 5. Geometric Modeling: a) 2D Wireframe b) 3D Wireframe c) 3D Surface Modeling d) 3D Solid modeling

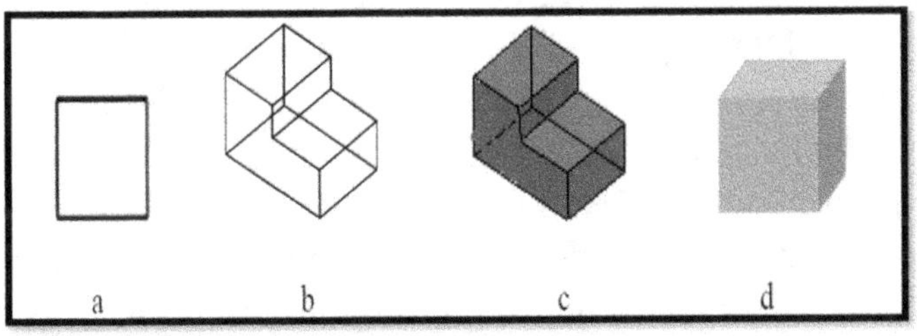

CAD Modelling

CAD Modelling is discussed separately in below sections.

Geometric Modelling (Vector)

Wireframe Modelling

There are several forms of modelling that are grouped under geometric vector modelling class of drawings in CAD. Wireframe models are one such. Wireframes have no surfaces. These have only edges which form the models. Wireframes are considered to be appropriate transition from 2D drawing mode to 3D. Wireframes are limited by lack of features to allow for inclusion of information on associated areas and volumes. Wireframes are also not ideal for presenting a simulated view of what an object would look in real life.

Surface Modelling

A step up from wireframe models, surface models have a coat or layer drawn over frame. Using complex algebraic and geometric equations, areas between lines and edges in wireframe model are drawn to produce a surface to cover object.

Solid Modelling

Solid models are considered to be easiest for users to perceive and comprehend. These models allow users to view life-like digitized version of the actual product from almost any angle.

Figure 5 shows the different forms of geometric models.

Object-based Model

In the last decade of the 20th century, object-based computer-aided designs began to be developed and used. One of the early examples of such model was Architectural Desktop or ADT, now known as AutoCAD Architecture. AutoCAD Architecture allows "intelligent" 3D modelling. Architectural objects such as walls, doors, and windows can be added to drawings. Because objects are "aware" of their design, form, function, and relationship to other components, object-based models produced by AutoCAD Architecture are considered to be intelligent rendering of engineering drawings and designs.

Constructing a physical structure involves three interlinked disciplines. The terms architecture and engineering are associated with design, while the actual process of construction describes the actual steps involved for product planning and production (Turk, 2000). Today, several specialized computer-aided design software available in the market offer user-friendly features for architecting columns, beams, windows, doors, floors, and roofs.

Parametric Modeling

The construction industry is considered integral or in certain cases it is viewed as a superset of Architecture Engineering Construction (AEC) industry. Information Technology offers intelligent geometrical features and databases together called "parametric" for the AEC industry as a whole. Parametric models use 3D to create digitized designs and drawings which are then stored in computer database systems.

Building Engineering Applications

Computer-aided design software products have progressed beyond standard designs and drawings. Software applications that are sold for engineers and architects can be found specializing in energy, HVAC designs, lighting simulations, and structural analysis to name a few. Such applications facilitate in exploring various designs in order to arrive at a desired optimum design solution. Another important advantage of these software is that designs can be created taking into account building regulations.

Computer Aided Estimating

Construction companies are often found to work on wafer thin margins. It, therefore, does not come as a surprise to observe that organizations in this sector view management and controlling of costs a vital factor to ensure survival in global markets that are often rocked by waves of recession and fierce competition. Construction companies must be able to make an initial determination of costs and bills of materials before bidding for tenders. Software products are available in the market that enable managers, planners, and engineers prepare accurate cost estimations and bills of quantities. Latest versions of cost estimation software are also able to exchange data with CAD software to provide an integrated solution. The obvious advantage to this is that data redundancy can be minimized, while calculating estimates quickly and as devoid of errors as possible.

The rapid pace of industrialization and emerging highly competitive markets across the globe have thrown into sharp relief the many disadvantages of using computing through manual means or even relying on simple software application. These disadvantages are listed below:

1. Processing time lengthens, often exponentially, with progressive complexity in designs and drawings.
2. Repeated manual entry of same data leads to errors or typographical errors.
3. Critical areas for consideration may be entirely overlooked by mistake or through lack of adequate information.
4. Information and data exchange may be stymied owing to lack of a cohesive structure for porting data among different cooperating agencies and departments.

Success in the construction industry is not only measured in terms of managing costs alone, but also in terms of safety, reliability, robustness and endurance. The latter factors are dependent to a very large degree on the accuracy of figures and input data. As such, it becomes a matter of keen interest for all concerned to actively look for tools and technologies that would not only facilitate in quicker and more responsive operations, but would offer all of that without eschewing accuracy and data fidelity.

Commercial Estimating Software

Commercial cost estimating software should be an indispensable tool in the quiver of any construction company. Not only does such software assist in processing data related to areas, volumes, and other complex numerical tasks, it also integrates with databases to extract information related to costs of materials, equipment, and of related human resources. The need to re-enter data in different standalone software is automatically negated when commercial cost estimating software are used.

Takeoff (Digitizer)

At the heart of cost estimation in construction business, is the take-off process. In the take-off process, the various components and elements that are required for physical construction are identified, measured, and quantified. Without this process, work on preparation of bills of materials cannot commence. Information Technology has altered the way take-off process is now handled. Instead of stationery and pens, computerized interactive digital systems have replaced scales and rulers. But such interactive digitized tools are not without their own set of drawbacks. These tools, though replacing pens and papers, can only produce 2-dimensional drawings. As a result, quantity surveyors are left with no other option than to compute third dimension manually prior to digitization process. The use of interactive digital systems also does not rule out duplication of inputs. These drawbacks have, therefore, led to levelling of criticisms at it from different quarters all claiming that it requires more time than manual approach to cost estimation.

Pricing

Assigning rates and prices to components and elements selected in take-off, is known as pricing process. The role that Information Technology plays at this stage is noteworthy. Unit prices and production rates only need to be stored in database for such to be assigned to products and components, accordingly, and systematically.

Summarization and Reporting

The use of estimating software has considerably facilitated in preparation of assorted reports. Information Technology considerably speeds up process of data amalgamation and allows generation of numerous reports quickly and accurately.

Spreadsheets

The emergence of spreadsheets in the latter years of the 20th century, rapidly revolutionized how extremely large numerical data are handled. No longer were sophisticated number-crunching software consigned to large dedicated computers – either mini- or main-frames. Desktop computers accorded a level of flexibility and processing power that was hitherto unheard of or unthinkable. Spreadsheet software began to rapidly evolve too, largely filling in the void those larger computers with their own set of dedicated software could not occupy owing to enormous costs to buy and maintain.

Soon, spreadsheet software on desktop PCs began to grow increasingly sophisticated. The price of sophistication was also accompanied by increase in such software costs. But large constructions companies have not shown reluctance in procuring such spreadsheet software programs. For one, these software cost fraction of the cost it would take to purchase a single estimating software, without sacrificing any of the benefits or ease of operations. Spreadsheets offer the following advantages:

1. They are usually comparably inexpensive to purchase outright, or even subscribe to on an annual or monthly basis.
2. Spreadsheets are user-friendly. The learning curve of operating a system is not unusually steep.
3. Spreadsheets allow for customization.
4. Notwithstanding visual appearance and ease of use, spreadsheets are incredibly powerful tools.

Among several free and paid spreadsheet solutions that are available in the market, none are more popular than versions of Microsoft's Excel. Offered in terms of perpetual license or on subscription basis, Microsoft Excel has a variety of spreadsheet functions and features that allow companies to automatically perform complex numerical analysis and data extraction on large inter-linked data. Indeed, Microsoft Excel can even pull data from enterprise database systems and present assorted forms of customized reports. Although, setting up a Microsoft Excel spreadsheet for estimating tasks can take up time, subsequent data processing becomes systematic and less time consuming.

CAD Integration With Estimating

Researchers are constantly on the lookout for improving models that enable use of CAD files for measurements and estimating. The International Alliance for Interoperability (IAI / IFC) regulate the standards of CAD integration. It has, however, been pointed out that the applicable learning curves for both designers and constructors have yet to be analysed in greater detail. Despite this, the benefits of an automated model to directly use CAD files for measurements and estimations are widely acknowledged as being significant in terms of cost and time savings.

PLANNING, SCHEDULING AND SITE MANAGEMENT

For years on end, construction industry relegated computers and information technology to somewhat of subordinate role. Even today, there are probably few around who would argue the primacy of information technology. On the contrary, computer systems can help act as a force-multiplier for personnel working at sites directly or in administrative roles at remote project locations. Web-based collaboration tools and software have considerably have had a positive influence on productivity at site level. As a step up, construction industries are now beginning to actively explore information technology tools to allow them to use mobile technology. Mobile technologies have allowed for abstraction of physical hardware layer and accelerating sharing of data over the Internet. No longer are users, who rely on mobile devices, tied down to their personal computers for viewing or sharing data. As long as stable and reasonably fast Internet speeds are available, a mobile device can easily upload files and allow for viewing in any geographical location. In trying to examine and explore the effect that Information Technology can have in the construction sector, an initiative titled “Construction Opportunities for Mobile IT” was launched in the United Kingdom.

Table 2. Comparison between Primavera and MS project

Microsoft Project	Oracle Primavera
Developed and marketed by Microsoft	Primavera software was the product of the eponymous Primavera Software, Inc. Considered to be one of the leading Project Portfolio Management (PPM) software of its class, the company was acquired by Oracle later on.
Users prefer MS Project for similar looking interface comparable to other Microsoft office suite software, and ease of use	Relative high learning curve.
Considerably cheaper to own	More expensive than its counterparts
MS Project allows a limited number of baselines	Offers unlimited baselines
Does not support real-time collaboration on a single project by multiple users	Supports real-time collaboration on a single project by multiple users
Does not offer creation of sub-activities	Offers creation of sub-activities
Supports up to 40 columns	Supports more than 200 columns
Cost-effective for use in small projects	Preferred in case of large construction projects

Table 3. Management Software used by Civil Engineers

Software	Brief Detail of Purpose
Project KickStart	Able to aid in identifying goals, challenges, and set up individual objectives for projects
AMS Real Time	Developed for assisting engineers create plans and monitor work objectives and resource utilization in real time.
Deltek Open Plan	Marketed as a tool that offers multi-project analysis, and is able to offer critical path planning in addition to resource management.
FastTrack Schedule	Projected as a user-friendly software that helps engineers carry out project planning and monitoring intuitively.
ZoHo Projects	A web-based application that supports task management, and preparation of Gantt charts.

Planning and Scheduling Software

Commercial Planning Software

A construction project involves countless number of man-hours devoted towards planning and management of both physical as well as human resources. Information Technology can help in formulating plans and monitor resource management at every step of a construction project. A number of planning software and tools are available in the market. However, the top tools acknowledged by industry experts are Microsoft Project and Oracle Primavera. Table 2 showing comparison of these two software has been provided to illustrate the key differences and functionalities.

In spite of the obvious advantages and disadvantages fielded by each of the two software, users continue to show preference for these above most other software currently available in the market. A summary of the several recognized project management software has been detailed in Table 3.

4D CAD (CAD Integration with Planning)

Construction planners and engineers have long sought to add the facet of time to digital designs. As a result, software products that offered 4D CAD, as such came to be known, began to appear in the market. Linking 3D graphical models to temporal aspect, 4D CAD models enable engineers to examine construction plans graphically and far more intuitively.

However, despite its obvious advantage, 4D CAD offers little by way of analytical features. The role of the planner has not diminished despite the advent of 4D CAD software. In many cases, 4D CAD augments the planner's ability to visually test whether problems can occur during construction process.

Development of 4D CAD models are labour-intensive and usually involve a number of steps. These are considered by many to be inherent limitations. These are as follows:

1. 4D CAD models are hobbled by the fact that not all information is displayed to help in evaluating current project schedule.
2. 4D CAD models lack the ability to inform users of constraints that affect sequencing of objects in drawings.

Figure 6. Overall System Architecture for 4D CAD Model

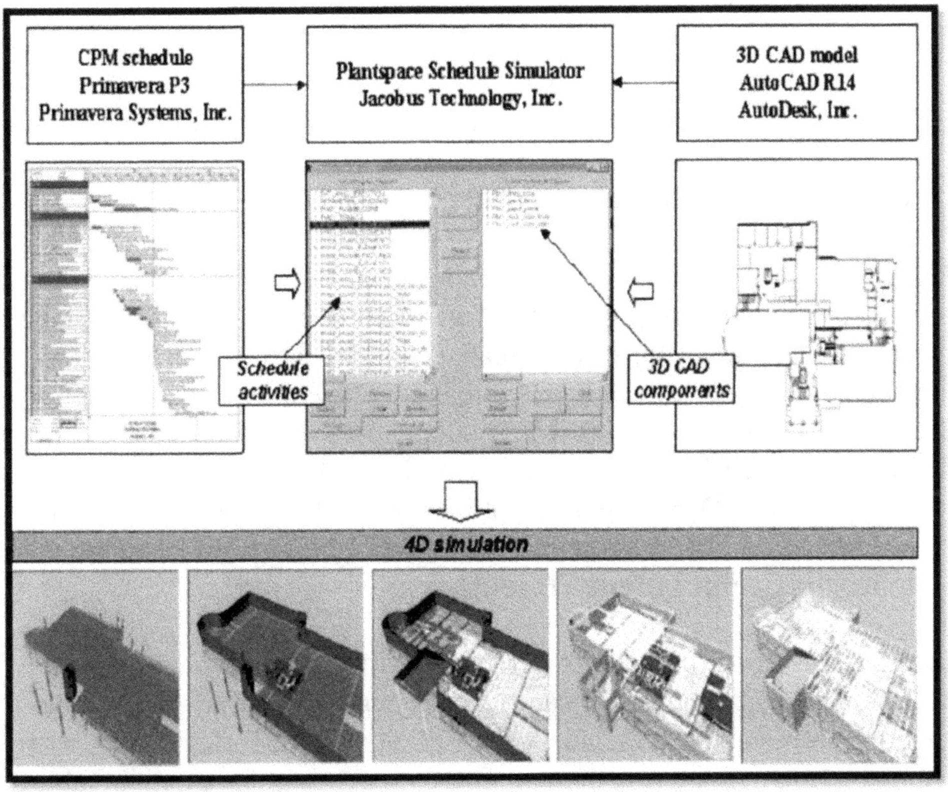

3. It has been argued that 4D models do not present all relevant planning data as per CPM. Further, 4D models lack the ability to alert users to the availability of float activities.
4. 4D models do not reveal activities that are missing corresponding components.
5. 4D models do not offer the flexibility of view a project from multiple perspectives.
6. Immediate viewing of alternative scenarios is not possible in many current versions of 4D software. In such cases where a change to the model or schedule is required, users need to consult the relevant 3D model of schedule, as the case may be, in separate designated applications.

It becomes abundantly clear that it would be far convenient and effect if construction schedule of a building is automatically produced from the drawings furnished at design phase. This has a two-fold benefit:

1. It would be easier to devise more accurate construction schedule at the design phase.
2. Business managers, project planners, and construction engineers would be in a position to thoroughly evaluate data and allocate appropriate human and material resources well in advance.

Computer Aided Facilities Management

The rise of Facilities Management (FM) in construction sector began approximately with the popularity of personal computers in 1980s. About this time, construction companies realized that maintenance is a vital factor essential to compute cost of a structure over its lifetime. FM software today is a combination of CAD mated to relational database management systems (RDBMSs). Since, FM software needs to harvest data from design drawings; the database (RDBMS) is significant in context of construction operations.

Monitoring Construction Projects

In a typical construction project comprising of 5 stages, Information Technology enables project planners and engineers to take advantage of superior processing power and data integration that assorted software tools can provide over manual approach. These aspects are discussed in Table 4.

Table 4. IT-Enabled Changes to Project Processes

Life Cycle Stage	Ways in Which Information Technology enables Processes
Understanding and Defining the Requirements	• IT tools offers key visualizations to take place. • GIS assists in examining physical locations for sites. • Facility Managers are able to actively participate in the initial stages using FM software. • As advancements in IT continue to take place, repositories of knowledge can be built and maintained for future use.
Process of Tendering and Team Selection	• IT tools already support E-tendering and E-businesses. • Advancements in software would make information sharing an extremely critical element in selection of teams for execution of projects. • Significant long-term business relationships can be created and maintained among all stakeholders in project.
Design Process	• IT tools that offer visualization features can boost communication. • Simulation software enables conducting analysis and carry out tests. • IT can significantly simplify and aid in reusing existing knowledgebase.
Construction Process	• Smartphone apps can be used to obtain visual data on different structural components throughout construction stage. • Tagging of materials and use of bar codes can help in systematic tracking and inventory management.

TOP DIGITAL TOOLS AND TECHNOLOGY USED IN CONSTRUCTION

Among the several important advancements being made in Information Technology, progress made in wireless systems and network promise significant improvements in the field of construction industry. While wireless itself is not new, the rapid strides made in mobile telephony, cloud computing, the Internet, and software development have coalesced to form a fabric that has already begun to yield benefits for construction companies who have adopted wireless network technology for data and information exchange.

Mobile Hardware

Even at the turn of this century, it was not uncommon for personal digital assistants (PDAs) to be used at construction sites. However, with the demise of PDA and dominance of smartphones, tablet PCs, wearable computers and digital hardhats, and digital pens that can be hooked to computers, construction companies are now able to reap the rewards of adopting Information Technology at ground level.

Personal Digital Assistants

Personal Digital Assistants that went from being promising technology at the turn of this century to being uprooted by Android and Apple smartphones, were sophisticated digital versions of diaries and organizers. At the very fundamental level, a PDA offered a calendar, task list, and memo pad. At the height of popularity, PDAs were either powered by Palm OS or Pocket PC operating systems. While the demise of the PDA hardly registered in the mind of the masses, the capability of syncing handheld devices with personal computers and laptops is what laid the groundwork for later mobile devices to widely appeal to the masses around the globe. Similarities between PDAs and modern smartphones are to be found in that both these types of devices have wireless connectivity, Bluetooth, and voice recognition capabilities. However, what sets mobile devices apart are more powerful processors, the ability to install and run multiple software or mobile apps, GPS, and the capability to collaborate and communicate over the Internet. Many smartphones today offer OCR, QR code scanning, and barcode scanning features straight out of the box.

Handheld Computer

Handheld computers are usually devices that are larger than PDAs and have similar kind of functionality as conventional laptops but only in a smaller package. Handheld computers feature QWERTY keyboards and have smaller screens. Again, these devices were elbowed out of the market by the arrival of Android and Apple smartphones that offered progressively more sophisticated features with processing and storage abilities that rivals standard personal computers and laptops.

Pen Tablet/Touch PC

Tablet PCs, also known as pen computers in some select markets, are so named because of touch screen interfaces that respond to human fingers or by specialized digital pens known as stylus. Most tablet PCs are similar to laptops and personal computers in terms of specifications and even run standard desktop operating systems. Tablet PCs fill in the gap between smartphones and personal computers, and have proved to be successful and popular for users working at remote locations away from any conventional office environment. A tablet PC may be no larger than an A4 paper, but with specifications comparable to mid-range laptops and desktop PCs.

Rugged Notebooks

Rugged notebooks are no more than standard laptops that have toughened shell casings and are manufactured in a way to withstand harsh weather and climatic conditions. Rugged notebooks are popular with users who wish to have serious processing power while roaming in all-terrain vehicles exposed to harsh elements. Rugged notebooks may be no larger than normal laptops or slightly larger given the added bulk of protective casings that enclose their vital innards.

Wearable Computers/Digital Hardhats

Wearable computers are considered to be among the latest evolution in personal computing. These are devices that a user can wear on one's shirt or jacket. These devices come with audio and video capturing features and are also able to transmit data across the Internet. The wireless attribute of wearable computers contributes largely to the ease, convenience and popularity of such devices among users who are required to work in remote regions and in the field.

Digital Pen

Of all the hardware discussed above, digital pens are considered a novelty by many. Capable of storing information running into several hundreds of sheet at a time, digital pens have been projected as a solving the problem of recording handwritten notes / memos as well as data entry.

Figure 7. Smart Devices

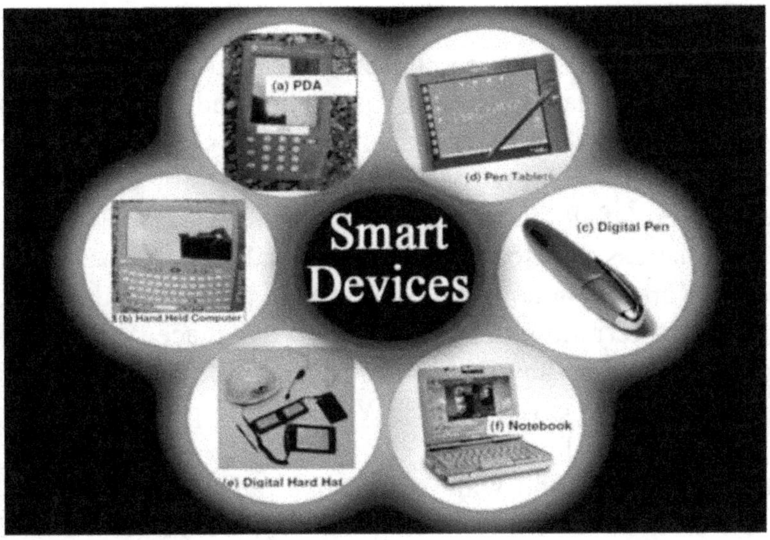

Networks

Information Communication Technology, today, offers a range of network types that have evolved based on user requirements. From the simple wired network through to intricate wireless wide area networks and beyond on to satellite networks, Information Communication Technology offers flexibility and operational ease for construction companies that run project sites at remote locations that can be almost anywhere on the face of the planet.

Wireless Wide Area Networks

Wireless Wide Area Networks (WWANs) are, by design, intended to cover more ground that Wireless Local Area Networks (WLANs). Already in most developed nations, and in many developing nations, 4G technology is now considered de facto standard or close to being one. It is now possible to transmit and move humongous amounts of data very quickly and, in certain cases, extremely cheaply. The popularity of 4G networks are in abundant evidence in India where the rates of Internet have fallen to levels where such no longer raises eyebrows – which once would have been the case only as recently as the turn of the century.

Wireless Local Area Networks

In construction projects where connectivity cannot be supported by wires for obvious reasons related to ease of installation and maintenance, Wireless Local Area Networks (WLANs) are the obvious choice. WLANs were developed across a wide range of industry standard protocol beginning with the IEEE 802.11. Of all the WLANs developed, WiFi is the best known and most popular across the globe. WiFi networks, however, are not known for covering large areas. For construction sites that are sprawled across acres of rugged or barren terrain with no infrastructure in place, WiMAX networks are more effective than any other form of network. With a range that may more than 40 Kms. in certain cases, WiMAX is appropriate in almost any complex infrastructural project.

Somewhat cheaper, and with an effective range that barely exceeds 10 meters at best, is Bluetooth technology. Today, alongside WiFi, Bluetooth technology is one of the top preferred medium of information exchange wirelessly. While WiMAX and Broadband Wireless require clear line-of-sight to operate, Bluetooth technology does not suffer from this limitation.

Satellites Networks

Satellite Networks are advantageous in regions which are beyond terrestrial network coverage such as mobile, WiFi, WiMAX, etc. networks. Initially expensive, satellite network services have progressively become cheaper. Satellite networks are eminently suitable for construction companies who operate a number of project sites that are beyond service coverage of other network technologies.

Mobile Applications

Mobile application developers are constantly researching and releasing newer versions of their mobile apps that can be used for CAD, data capture and retrieval, and project management. Mobile applications offer ease of use and flexibility in real-time and under any conditions provided Internet services are available.

CAD Applications

Users at construction sites are no longer required to be tied down to their desktops and laptops for viewing CAD drawings. Certain commercial mobile applications available in the market today allow mobile users to not only view 2D / 3D drawings, but also carry out modifications to a certain degree.

Data Capture Applications

The earliest forms of data capture used punch lists, time sheets, pens and clipboards at a time when micro-, mini-, and main-frame computers were the only known computers on the international information technology landscape. Today data capture techniques often rely on smartphones to collect real-time data such as those related to progress on piling activities.

Project Management Applications

Advancements made in Information Technology today have made it invaluable for users to rely completely on software that enable users to plan, monitor, review, and conduct risk assessments for a given project. There are several software tools presently in vogue such as Microsoft Project and Oracle Primavera.

CONSTRUCTION PROCESS WITH INFORMATION TECHNOLOGY

By virtue of it being a complex process in its entirety, construction process itself is segregated into sub-processes. Building lifecycle is perceived by construction managers as mostly a composition of two distinct classes of processes. The first being material processes, and the other information processes. These are illustrated in the figure shown below.

Material Processes

Materials processes are integral to materials and to the construction process as a whole. The quantum and specifications of the materials to be used, however, are controlled by information processes that include design specifications.

Information Processes

In the construction industry, information processes now have a stellar role to play. Project management teams rely extensively on information processes to design, build, plan, and monitor project details. As complexity in construction project increases, so too does the demand for sophisticated information technology rise proportionately. The costs of injecting more information technology and tools are usually offset by the speed and efficiency gained in managing a construction project.

Figure 8. Construction Process

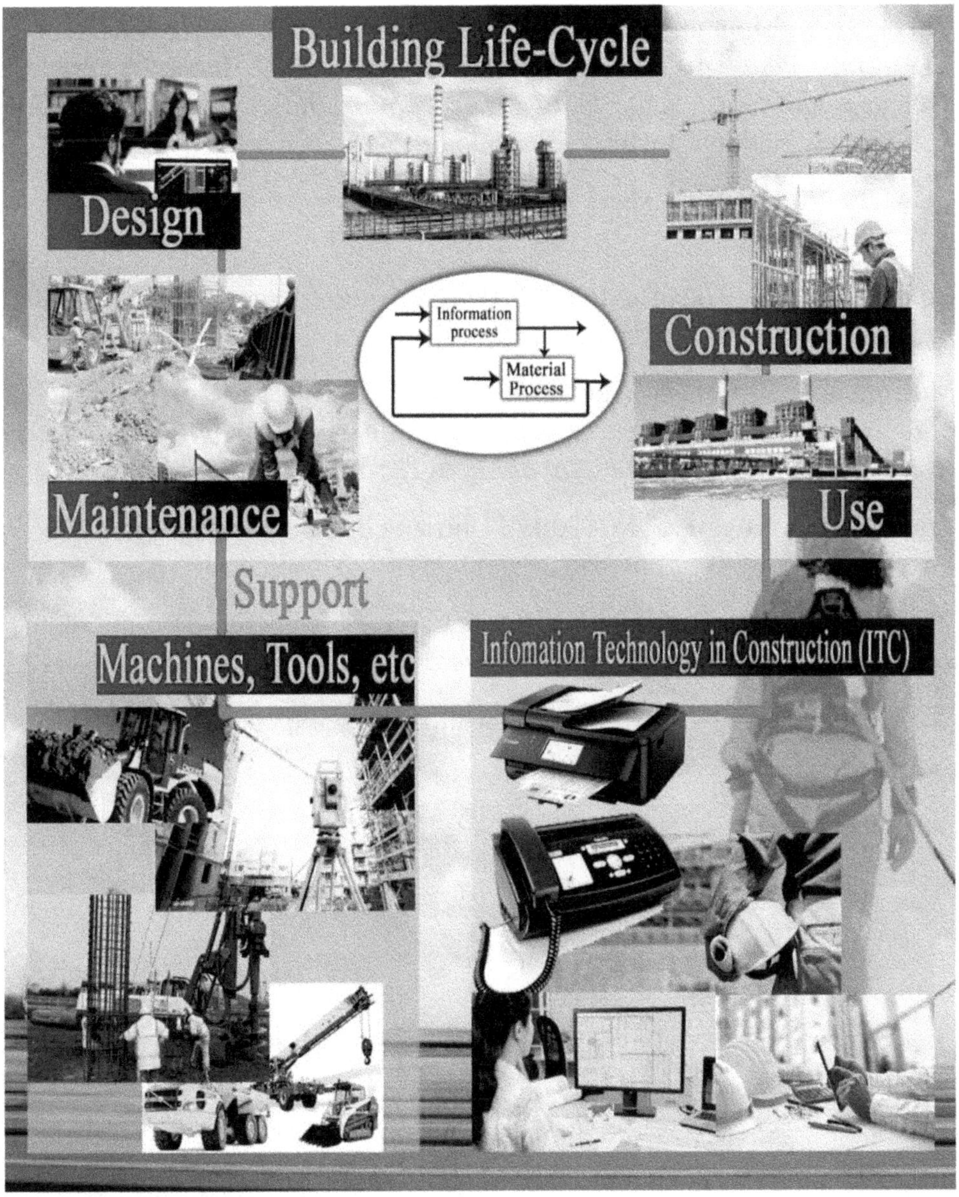

The use of Information Technology can significantly enhance and aid in every step involved in a construction project. From planning, designing, building, and maintaining, construction companies can reap rewards of injecting Information Technology at every step and phase.

REFERENCES

Cheok, G. S., Lipman, R. R., Witzgall, C., Bernal, J., & Stone, W. C. (2000). *NIST Construction Automation Program Rep. No: 4 Non- Intrusive Scanning Technologyfor Construction Status Determination, Building and Fire Research Laboratory*. National Institute of Standards and Technology.

COMIT. (2005). *Construction Opportunities for Mobile IT*. http://www.comitproject.org.uk/

Liu, L. Y. (1997). Construction field data collection using the digital hardhat. *Proceeding of Fifth Construction Congress*.

McCullouch, B. (1997). Automating field data collection on construction organizations. *Proc., 5th Construction Congress: Managing Engineered Construction in Expanding Global Markets*, 957–963.

Prasad,, R., & Ruggieri,, M. (2003). Unified Global Infrastructure. *Wireless Personal Communications*, *26*(2–3), 89–91.

Thorpe, T., & Mead, S. (2001). Project-specific web sites: Friend or foe? *Journal of Construction Engineering and Management*, *127*(5), 406–413.

Wikipedia. (2006). *WiMax*. Available at https://en.wikipedia.org/wiki/802.16e)

Chapter 2
Paving the Way for Better Construction Designs Aided by Information Technology Tools

ABSTRACT

With the dawn of the industrial age, construction and manufacturing methods have witnessed revolutionary engineering practices and methods. None have been more significant than that of the influence that information technology has had on civil construction engineering as in many other engineering disciplines and spheres. In this chapter, the authors discuss how computer-aided design and drawing software, the internet, building information modelling, and 4D and 5D CAD/CAM systems have revolutionized the way construction engineers, designers, and architects work. They also discuss the various types of software available in the market, each with a specific goal and engineering purpose. They also offer brief comparisons on several types of design software so that readers can grasp quickly the salient features that each have to offer.

Manufacturing methods have long since been influenced by developments in Information Technology. Especially since the 1970s and the 1980s, computer-aided design (CAD) and computer-aided manufacturing (CAM) systems have had a strong influence in manufacturing sectors. It was not long before companies began rethinking their manufacturing processes and redesigning these with the help of Information Technology. While manufacturing earlier

DOI: 10.4018/978-1-7998-5291-9.ch002

involved tedious and often expensively complex processes to create working prototypes before moving on to actual production, with the help of Information Technology and tools today it is possible to simulate actual working model in 3D. Designs can now be created, tested, evaluated and modified all in digital space. The concept of Virtual Design and Construction enables engineers to build and test their designs in virtual space. This allows them to evaluate several designs and explore alternative routes to match stated objectives as defined by contracts. Virtual modelling and simulation become invaluable and indispensable as complexity of a construction project grows. Architectural designs require complex collaborative effort from all stakeholders. Use of digital simulation and virtual modelling are influence design process and help in higher productivity and throughput.

Rapid strides being made in Information Technology has now made it possible for improving business process and rethinking strategies. While traditional CAD software could earlier present information in spatial relationships, domain-specific information would usually be absent. With the help of advancements made in Information Technology, it is also possible to simulate thermal behaviour in addition to structural designs. Building Information Model (BIM) is a stellar example of building a database of information that includes a wide range of parameters with the goal of sharing information beginning from architects, engineers, contractors, and suppliers.

INFORMATION TECHNOLOGY AND DESIGN

Advancements made in the Information Technology constantly influences the manner in which business processes are structured and followed across industry verticals. Construction companies have been known to be reluctant in adopting change. Even when computers were being widely adopted by other industries and companies engaged in infrastructural activities, the construction sector was slow to embrace computerization. Computer-aided design software were already in circulation in other engineering branches before architects and civil engineers caught on to the idea of involving computers in their profession. Initially, computer-aided design (CAD) software began to be used by architects to transform their engineering drawings from paper to digital. As CAD software evolved, architects and civil engineers began to explore additional possibilities boosted by the knowledge and experience of their colleagues in aerospace engineering, mechanical engineering, etc. With the advent of 4D CAD modelling techniques, architects and civil engineers could now visualize and test changes in design. For such obvious reasons, 4D CAD has been garnering far wider following in industry circles than either 2D or 3D models in recent times. Compared to traditional forms of planning, visual models created using 4D CAD software has been observed to be more accurate and of more use. However, not long before 4D CAD modelling began to be used widely in industry circles, computers and Information Technology were commonly used mainly for communication and as data processing tools. However, as the pace of Information Technology picked up, construction companies began to adopt it for knowledge integration and for supporting decisions in collaborative efforts involving complex projects. This, in turn, led to realization among construction companies that Information Technology can be used to rethink and redesign business processes. The introduction and popularity of high-level programming languages, viz., Java, Python, etc. began opening up new opportunities whereby construction companies could remodel their entire business approach.

The organization of process related activities changed, and so did products, after digitized architectural drawings began to be introduced. The following figure 1 depicts differences between practices that had been the convention in the past, and those that are now understood to be future. In the past, processes of manufacturing were set apart from the data that influenced final production. Today, processes are driven by data and information that are analysed well before any actual production takes place. However, the traditional approach has not yet been relegated to the back seat. Both old and new processes are mixed to produce a transitional stage wherein design processes are coalesced with knowledge and data received from various points of organization in order to fulfil contractual obligations.

Figure 1. The Role of Information then and as it is now.

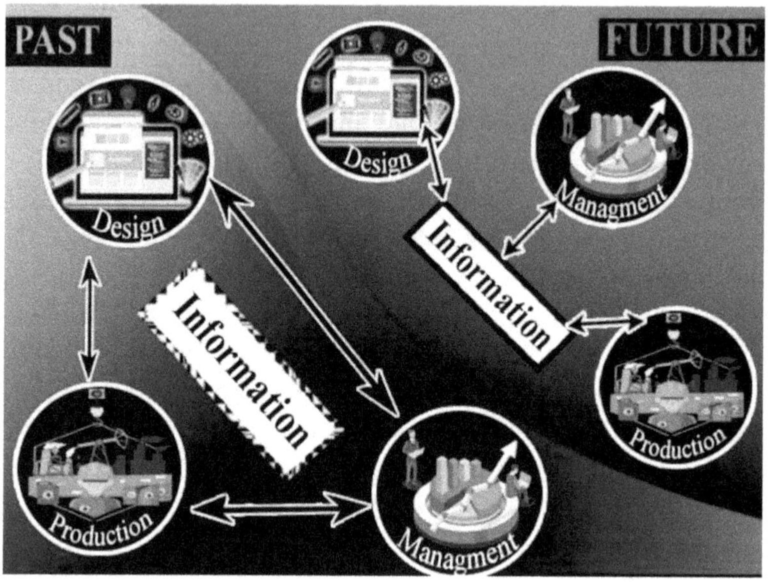

COMPUTER AIDED DESIGN AND VISUALIZATION

The advancements made in Information Technology that began with the invention of personal computers and, thereafter, with the rapid rise in operating systems and smartphone technology, have seen computer-aided design (CAD) software gain widespread popularity among the fraternity of architects, engineers, design professionals, and project planners. Among the various CAD software is available in the market, the more popular ones are developed by Autodesk, Bentley, etc. The modern CAD software has long since replaced traditional drawing boards at the production stage.

The primary function of any architectural or engineering group is to produce drawings quickly and efficiently, keeping in mind all structural safety and aesthetic requirements. Similar to CAD applications in any other engineering discipline, the use of CAD software in construction sector differs only in the manner of design and application processes. CAD software, today, are developed to be feature-rich insofar as design construction techniques, storing and sharing of drawing data in databases, and produce digital outputs that enable users to actual view and test designs even before a single concrete production step is taken. The advantage of CAD software over traditional drawing board methods is that the data and information related to drawings can be disseminated and edited across compatible application software and over a range of equally compatible hardware. CAD software also enables users to produced schematic drawings displaying accurate measurements and relationships between components in designs of structures that are to be produced. Further, drawings can be rapidly produced, edited and shared using CAD software. With modern CAD software, the metrics and parameters involved that define and can be used to reproduce drawings in another CAD environment, can be stored in databases. This feature holds the key to compute well in advance estimated material costs involved in production. Advanced CAD software can even produce life-like images along with interactive animation features that enable users to move through complex spaces and structures as if those were real and tangible.

AutoCAD Software

By far, AutoCAD remains the foremost choice for those looking for computer-aided design software on PC / laptop platform. The software is capable of producing any drawing that is possible using traditional drawing means using paper and drawing boards. 4D CAD software can even go to the extent of providing a spatial and temporal view for users to help in planning and mapping entire construction processes of even the most complex civil engineering projects of any magnitude. This is considered widely by many in engineering disciplines to be a total game-changing paradigm that separates 4D CAD modelling software from their 2D counterparts. CAD software is mostly preferred even at the basic level of engineering drawings over traditional methods, because of their ability to create, share, and edit drawing information among users with rapid ease and convenience. As such, experts often consider CAD drawings as a database of information. AutoCAD, which was first launched in 1982 and has since seen considerable changes and development to its core software code, is now considered (Akbas, 2004) the byword in any engineering discipline. Today, complex and intricate drawings requirements almost universally mandate use of some version of AutoCAD over any other CAD software currently in the market. Users prefer AutoCAD because of the myriad features that the versions of this software have allowed that multiply productivity, enhance speed and accuracy, and reduce errors while drafting. Not surprisingly, the 3D versions coupled with cloud-based connectivity offerings, have turned this software (Akinci, Fischer, and Kunz, 2002) into a major hit in any manufacturing sector. Aside from Autodesk's AutoCAD, there are several other equally renowned vendors who offer software that have Building Information Modelling (BIM) workflows as well. Following is a brief overview on software that has made a mark in computer-aided designs.

Table 1. Drafting Software Used by Civil Engineers

Software Name	Description
Vectorworks	First designed and developed by Nemetschek Vectorworks, Inc. in 1985, it offers BIM in addition to 2D / 3D modelling, and various additional features.
Autodesk Maya	First developed by Alias Systems Corporation and later own purchased outright by Autodesk, Inc., Maya remains the world's leading choice for 3D computer graphics including video games, animation, etc. Civil construction engineers depend on Maya for their graphically-rich design and modelling requirements.
Rhinoceros (Rhino)	Touted as world's best-in-class easy to use design software, Rhino is a 3D CAD software that was developed by Robert McNeel and Associates.
SketchUp	Google's SketchUp was offered as a free CAD software at one time. Today, the brand is owned by Trimble and is offered at different price points for both home as well as corporate users.
FreeCAD	As its name suggests, FreeCAD is among the first among open-source CAD software applications that allow users to customize the very source code itself. Targeted at users looking for reliable open-source application that supports 3D modelling as well as script support, the software has gained traction among users in architectural and engineering fields.
LibreCAD	LibreCAD is an open-source 2D CAD application that remains a viable choice for users who look no further than 2D design drawing requirements.
nanoCAD	Marketed as a competitively priced 2D CAD software tool, it provides an attractive interface and native DWG file support.

INTERNET-BASED COLLABORATIVE DESIGN

Complex engineering construction projects often involve diverse sets of teams working in partnership with each other. More often than not, teams may use different CAD tools to match their own exacting design requirements. However, this throws up a problem involving consistency issues while sharing assorted file formats produced. To obviate such problems, and given the fact that Internet is now widespread, organizations now prefer CAD software (Amor, and Faraj, 2001) that support collaboration and sharing of design-related data accessible from any remote corner of the globe with Internet connectivity. It has been opined that for a CAD tool to be truly considered collaborative and serious contender in any major engineering organization, it needs to correctly identify and allow working on CAD formats ranging (Andersson, Suber, 2006) from the esoteric to the popular ones widely adopted in the industry, support real-time communication, allow publishing 2D / 3D CAD on the Internet, and allow editing CAD models developed in respective native CAD software.

Real-Time Communication

CAD software developers have banked heavily on Internet to support real-time collaboration of files. Modern CAD software now enable designers to utilize features for connecting over remote sessions using the Internet to work on any kind of designs or models on their respective applications.

Support for Various CAD Formats

To effectively support real-time collaboration, CAD software developers have long since realized the need to support vendor-agnostic file formats. Thus, organizations are able to work in mixed CAD software environments and access collaborative spaces without concerning themselves with complicated conversion processes. In turn, such support for vendor-neutral CAD software translates to a boost in productivity and is considered to be a force-multiplier in situations involving multiple stakeholders working in a complex civil engineering project.

Tools for Publishing and Viewing 2D/3D CAD Designs on the Web

While the popularity of sophisticated CAD software has never been called into question, there is nevertheless a very large market that cannot afford to buy and use these products. Often a project can involve vendors, consultants, and other contributing partners who neither possess requisite levels of CAD expertise nor can they afford to own and operate such products. In view of this, CAD software developers have responded by integrating a feature that allows to publish designs directly on web pages. This enables users to preview designs and understand key features without the need to actually install and operate complex CAD tools.

Tools for Manipulating CAD Models Outside the Original CAD System

Although, publishing designs on web sufficed for some, considerable number of users wanted a feature that would allow them to do more without breaking the bank and investing in CAD software. This led to CAD software developers to produce tools that offered users more than simply to view designs. Today, lightweight tools are constantly developed and marketed to enable users to carry out basic transformation, zooming and panning, and mark-up activities on CAD design software. A table showing salient features of such tools are given in Table 2.

Table 2. Features provided in lightweight CAD tools by CAD software developers

CAD Tool	Can Publish on Web	Can Add Notes and Markup
Autodesk Volo View	Yes	Yes
Solid Concepts Solid View	Limited	Yes
Parametric Product View	Yes	Yes

Building Information Modeling (BIM)

Although Building Information Modelling (BIM) has been trending in the construction industry in recent years, the fundamental concepts of this have been in existence for over several decades. In simple terms, BIM is an integrated information model that contains drawings, specifications of materials, details of constructions, and other paraphernalia housed in a centralized or distributed software database systems. The arrival of BIM in mainstream construction engineering has initiated a revision in design procedures and documentation.

Building descriptions typically include building systems involving electrical, HVAC, mechanical and plumbing components. BIM is reliant on Information Technology for management of data and allied processes. BIM is founded on the premise of virtual project formulation first so that project planning and management framework can be put in place well before actual construction commences. BIM allows virtual integration and design in 3D. While this is certainly a definite advantage, it is not the only one by far that BIM offers. BIM enables construction teams to collaborate and analyse designs very early on in the construction phases, and continually help designers, engineers, and architects in managing construction processes. Under field conditions, information on project sites and prevailing conditions can be fed into BIM and detailed analysis performed on such data sets. BIM offers 4D modelling in which entire schedule of construction can be plugged in with the actual model of the construction project to help in visually identifying construction activity sequences. Often referred to as 5D, cost estimation adds another important dimension that incorporates quantities and rates of materials involved in the construction project.

However, this is not all that 5D is all about. With the aid of this feature, it is possible to store such information in database systems and produce meaningful reports that can help in making pertinent financial decisions based on either perceived or computed economic factors.

Figure 2. BIM utilization during design and operation.

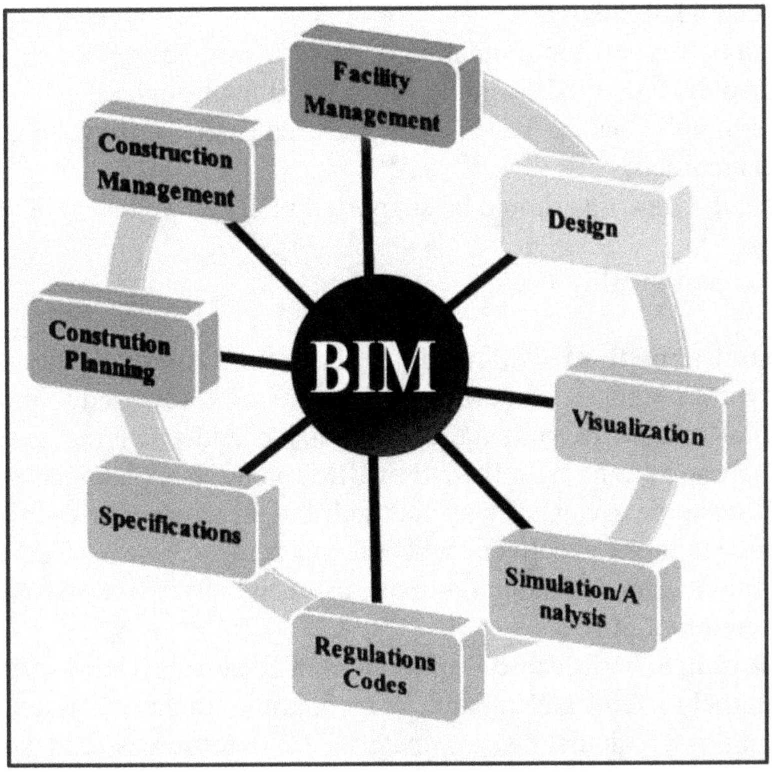

The primary objective of using BIM is to provide a common framework for information exchange among all stakeholders of a project. Further, BIM extends a step forward for Facilities Management wherein the health and structure of a construction can be closely monitored long after it has been constructed and commissioned. As seen in the Figure 2, BIM supports analyses and simulations to be performed to gauge structural integrity and efficiency of design.

One of the unique advantages of deploying BIM is that documentation associated to construction processes are automatically generated and updates to such are made in response to any change made to respective models. Data can be shared from BIM for consumption by external applications.

For BIM to be relevant for an organization searching for an integrated approach to management of complex engineering projects, the following features should be available:

1. Accessible: Data should be accessible.
2. Comprehensive: Design specifications including purpose and applications should be defined.
3. Digital: Support for simulation of construction design.
4. Durable: Data in BIM should remain usable throughout.
5. Measurable: Data in BIM can be retrieved easily, and can be annotated and recorded.
6. Spatial: Drawings should be supported in 2D as well as in 3D formats.

Application of BIM

- **Visual depiction:** BIM facilitates in intuitive visualization of even the most complex 3D designs. Projectors are often used to project 3D digital images to generate an immersive and engaging experience. Synchronization: With the aid of BIM, architects, planners, engineers and customers can all stay focused and in sync with each other to adhere to targeted delivery schedule of a project. Further BIM can help highlight errors or anomalies that can creep in from time to time during execution of projects.
- **Planning Schedule and Deployment of Resources:** BIM software are extremely useful and reliable in estimating conditions at actual field conditions, calculating raw materials and resources, and keeping a close watch on resource utilization.
- **Setting up Sequences:** Time is a critical factor that has to be considered in any project. This can be integrated into BIM to visually analyse

and take necessary optimization steps for preparing schedules and sequences throughout construction stages.

- **Computing Costs:** Quantities and rates can be stored in databases used by BIM. In turn, BIM can be used to ascertain costs and effect design changes to optimize expenses both in the short term as well as in the long run.
- **Incorporating Subcontractors and Suppliers:** BIM allows to directly pull in specifications of raw materials provided by subcontractors and suppliers. This reduces time involved, and greatly improves productivity.
- **Systematic Approach:** Using BIM it is possible to integrate all raw materials and resources under one umbrella. Equipment and their respective designs can be checked thoroughly, and conflicting areas can be resolved and addressed quickly.

BIM Levels Explained

BIM Level 1: Most organizations adopt BIM at this level. Mixing 3D CAD for conceptual drawings with 2D drawings for product information, BIM helps organizations meet statutory requirements. CAD standards that are BS 1192:2007 compliant are often set in BIM in Indian construction sector are considered, and relevant data shared. At this level, models are not shared between participating members in BIM.

BIM Level 2: At this level, each stakeholder of a project may have their own choice of 3D CAD models. However, a common shared model may not exist. Information is shared across common file formats that each participating member is free to perform their own respective analyses. The only mandatory requirement is that CAD software used should be able to produce outputs in common shareable file formats.

BIM Level 3: Also known as the "Open BIM", this level allows all stakeholders to access and share the same model. In other words, a conflict layer that might otherwise exist is done away altogether in this form of BIM.

ROLE OF INFORMATION: APPLICATIONS

In the 21st century, Information Technology and tools have made it possible for vastly improved project planning, visualization and simulation. Geographic Information System (GIS) technology and allied tools feature in most planning and development endeavours. Sophisticated graphical interfaces combined with GIS technology helps in producing almost life-like digital simulation of real construction projects in urban settings. No longer are planners hobbled by lack of adequate visualization in understanding spaces or lack of such within any infrastructure setup.

Figure 3. Virtual Design

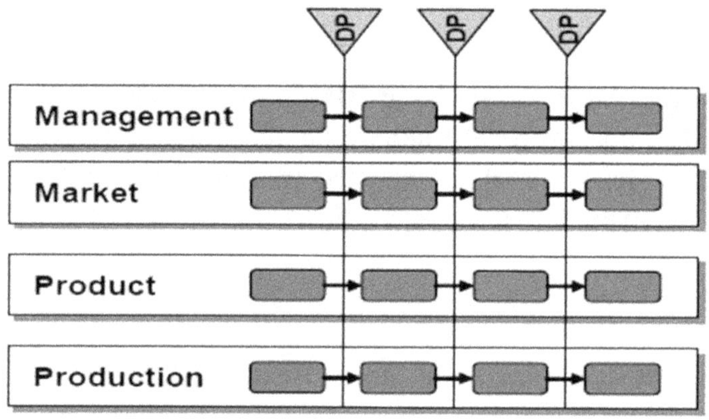

Experts have rightly predicted that BIM-based CAD coupled with latest in Information Technology will change the way information shall be processed, delivered and consumed. BIM facilitates in disseminating information with representation of features, effects on designs, and details of relationships that existing between components in a heterogeneous environment. Often BIM and product modelling are considered to be closely related to Virtual Design and Construction techniques. However, experts and advanced users disagree on this as there are specific construction techniques that do not fit the mould of Virtual Design and Construction.

Figure 4. Schema of a decision making process involved in a real estate project. Dotted lines depict primary decisions and the relationships with simultaneous related processes.

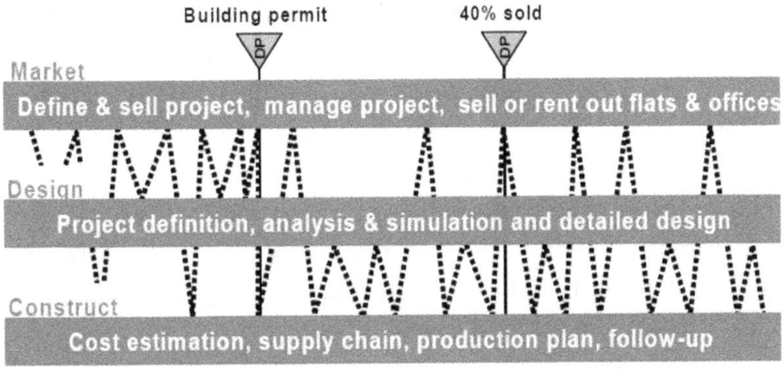

Figure 5. Pictorial depiction of Virtual Design and Construction techniques in a real-estate project.

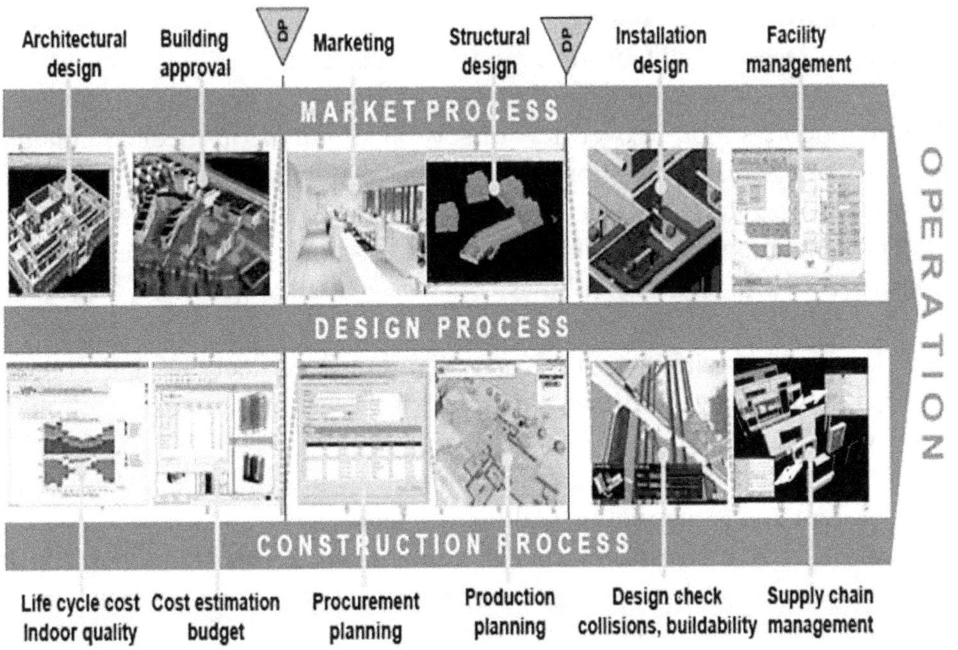

VIRTUAL DESIGN AND CONSTRUCTION

The purpose of Virtual Design and Construction process is to help analyse and predict quality of construction of a building or any other complex project. Virtual Design and Construction does so through simulation. For Virtual Design and Construction to be effective, process and the final construction project must be designed and simulated first in a virtual environment. This allows users to evaluate assorted designs and match these with the end objectives. While Virtual Design and Construction process relies extensively on CAD and related Information Technology and tools, it also utilizes disparate associated technologies to achieve end results. In India, Building Information Modelling is known as Virtual Design and Construction.

However, adoption of Virtual Design and Construction methods not only demand inclusion of appropriate Information Technology tools, but also mandates a renewed outlook at evaluating business processes involved. While Virtual Design and Construction methods have proven to be useful in a sequential product manufacturing environment, the actual benefits lie in using it in a development environment that supports simultaneous activities as seen in the Figure 3 below.

Virtual Design and Construction techniques can be used in a number of varied applications within an integrated building process. The following Figure 4 depicts a basic overview of several simultaneous decision making processes in action. The subsequent Figure 5 displays Virtual Design and Construction techniques for accelerating designs and corresponding processes.

The advantages of Virtual Design and Construction techniques in a civil construction project are as follows:

- Digitized models aid in clearly defining specifications and create a path for engaging in systematic communication among stakeholders involved.
- Authorities often issue permits quicker as a result of efficient Visual Design and Construction techniques adopted.
- In case of real estate projects, suitable marketing strategies can be adopted to fetch better rates based on feedback received from virtual representation of the finished project.
- It becomes considerably easier to change designs to meet stated objectives.

- Virtual Design and Construction makes is easier to plan procurement and delivery schedules that may often translate in lowering production costs to take advantage of market and rate fluctuations.
- As a result of collaboration that is transparent and systematic, chances of confusion and complications are reduced. This in turn leads to lowering of re-work on projects resulting in closer adhesion to timelines.
- Work flow and construction are significantly improved as a result of 4D approach to design and productivity.
- Wastage and excessive use of raw materials can be reduced
- As a result of improved facilities management, ownership value increases.

PROJECT PLANNING USING 3D AND 4D CAD.

The role of construction managers is to closely monitor and engage in activities to ensure that a project design is implemented successfully from concept to real life. At every juncture, construction managers need to keep a close tab on resources, budget and timelines. As a consequence, project planning is considered to be an extremely vital part in any construction project. Project planning requires paying close attention to designs, task, construction techniques to be employed, coordination among various stakeholders, and supply and allocation of resources. For a production plan to be effective, it has to cater to cost control, supply chain management, visualization and simulation of designs, and resource planning at every stage. In any planning process, Work Breakdown Schedule takes precedence. A Work Breakdown Schedule (WBS) provides an illustrative view of all elements involved in a project and corresponding relationships for a project to move forward. Scheduling and carrying out cost estimates are possible with the help of WBS.

In a construction project, planning process is gradually developed through a series of phases:

The parameters, requirements, and associated functions are initially set at project conceptualization stage. It is in this stage, that preliminary estimates are prepared considering generous margins to accommodate any unforeseen expenses.

In the next stage, master plans are drawn up to present an overall view of a project in terms of time and costs. It is at this stage that components for the various sub-structural components are identified, list of manpower drawn up, and sub-contracting details for the actual construction are prepared.

Figure 6. Creating a 3D production model through mapping of WB to 3D design model

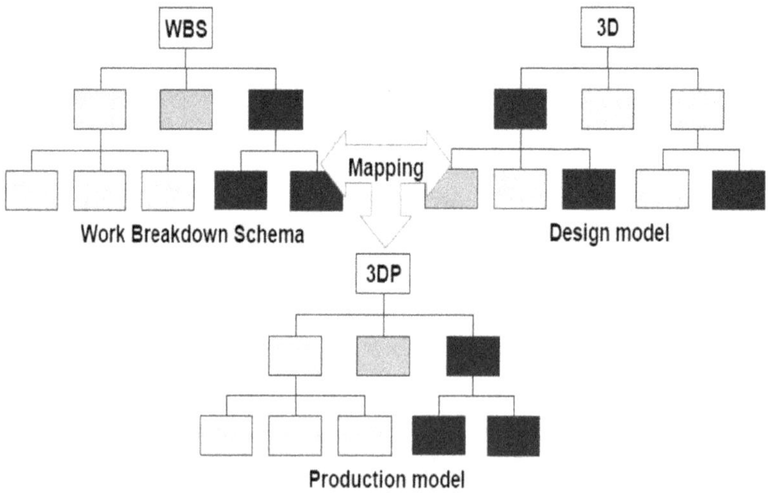

In the subsequent stage, actual designs are prepared and construction work starts. Scheduled work plans are constantly consulted to ensure that resources and deliverables proceed as per plans.

All the above stages require involving WBS at different appropriate levels. Without WBS, scheduling, estimation, planning and monitoring is not only impractical, but also detrimental to the very business of construction.

WBS, QUANTITY TAKE-OFF, COST ESTIMATION AND SCHEDULING

The purpose of a WBS is to provide a single reference umbrella for a project. Segregated into a hierarchical form of structure, WBS allows users to work on individual components that are also otherwise referred to as work packages in industry parlance. What WBS enables users to do is to load cost estimations and resource schedules at various stages and at assorted level of details. WBS makes it possible to identify those project sections that can be allotted to sub-contractors.

It is common for building artefacts to be designed and arranged in an object-oriented structure within 3D model. However, 3D models that are constructed during design process are not commonly found to correspond with WBS of related project. It, therefore, becomes of vital significance to connect the 3D objects to the corresponding work packages in WBS.

Figure 7. The figure above illustrates how 3D production model aids in developing cost estimation and scheduling processes. A work package consisting of activity description, task location, and resources involved is created using 3D models.

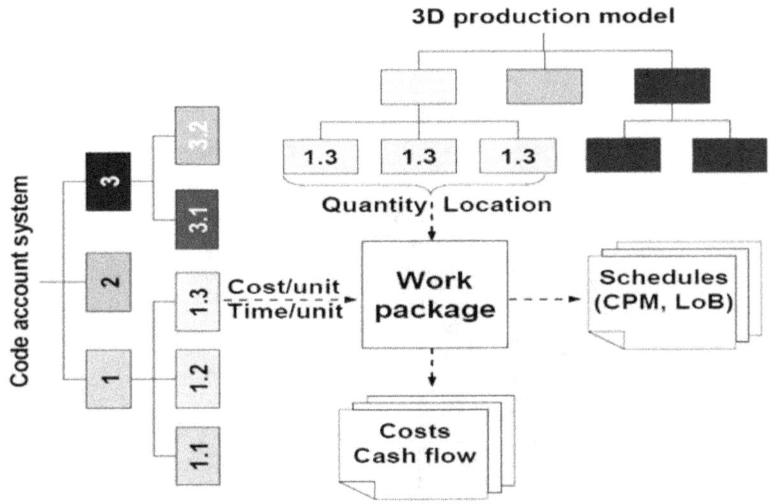

As the first step to organizing 3D models, a WBS of a project needs to be associated with the relevant 3D object models. An example of this is shown in the following Figure 6. Most users approach this using a technique wherein objects are mapped to the identified WBS work packages through additions of 3D objects to CAD layers that represent required level in WBS. The resultant model – that, at this stage, is called production mode – is now in a position to be modified further through addition of objects such as form work, shoring, etc.

Constituents of work packages in production models are defined in the next stage. The process is depicted in the following Figure 7 and Figure 8. Usually a work package is an amalgam of a number of tasks, or methods, that is intended to perform a set activity. As an example, constituents or recipe for a concrete column can include methods such as “tie reinforcement”, “cast concrete”, etc. Each activity would have relevant units of measurements associated.

At the organization level, each activity is allotted a unique code which is used to define raw material costs and work involved, along with man-hours required per unit of activity to achieve production status for an associated building component. Planning processes are usually supported by digitized versions of codes and data for 3D objects that are correspondingly stored and maintained. These are done via combination of the following:

Figure 8. Often to be found referred to as "recipes", tasks grouped with details on resources and requirements are mapped to 3D objects.
Illustration by (Graphisoft 2006)

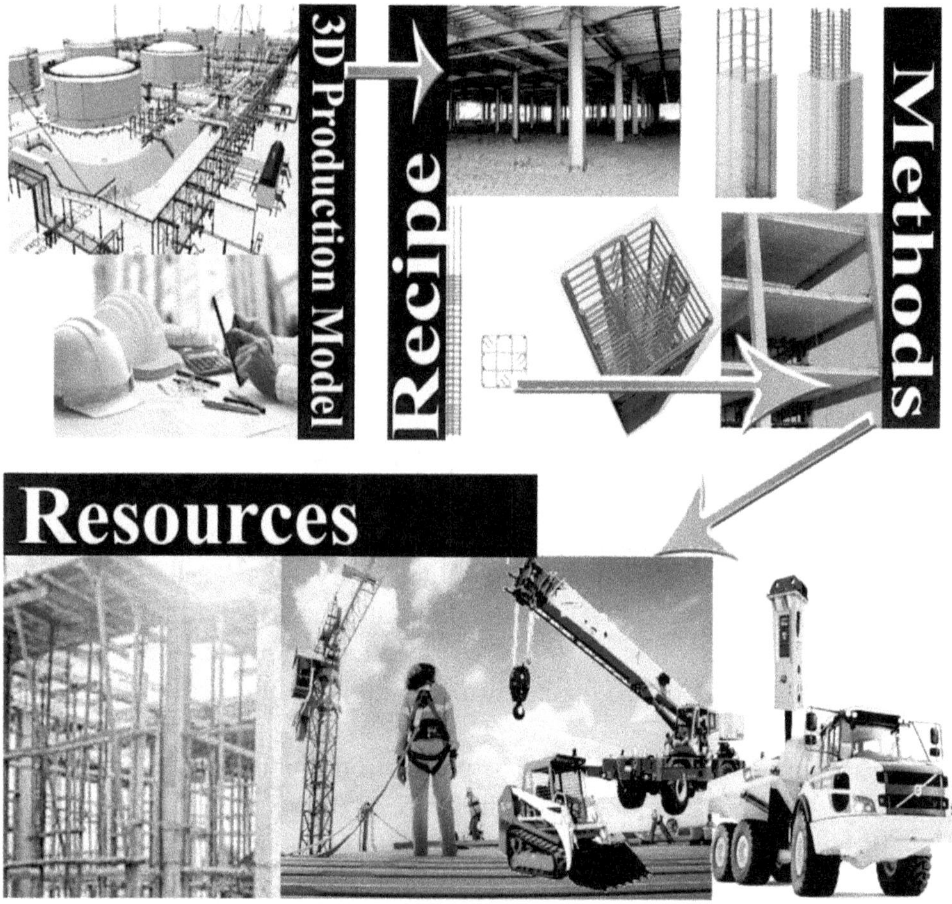

Preparing quantity take-offs using 3D models where estimation of costs can be built up from database containing data on building components. At any level of detail, construction costs of a project can be summarized.

Preparing quantity take-offs extracted from data contained in 3D model wherein details of building components are used for estimating time involved. The resultant estimates of time for given locations are used to schedule work, manage supply chain of resources, involve design information and share relevant details, and control the flow of cash throughout construction stages of a project.

Scheduling Methods

It is well understood and known that appropriate scheduling can eliminate any obstruction during project execution. Elimination of any potential causes of bottlenecks is a sure-fire way to ensure unimpeded progress during execution of various activities in a project. However, the task of scheduling itself can be a serious contender for the status of the most difficult of all hurdles that managers and project planners face. During the planning stages, managers and planners need to carefully consider and weigh the myriad of factors that can affect productivity of production crews and availability of resources. WBS helps in breaking down all the activities that make up a project into seamless and yet segregated schedules that consider availability of resources, manpower and equipment at any given point of time. One of the more important aspects of WBS is to pay attention to time and space continuum while arranging the buffers between inter-related activities. In subsequent sections, we approach the two well-known methods of construction project scheduling, viz., scheduling based on activities and scheduling based on locations.

Activity-Based Scheduling

The Critical Path Method (CPM) is an activity-based scheduling method that is widely applied in construction projects. Activity-based scheduling gained widespread acceptance with the adoption of pre-fabricated structure manufacturing and complex assembly processes. Considering the demand for this form of activity based-scheduling, Information Technology today offers sophisticated software that enable users to develop CPM-based schedules quickly and relatively conveniently.

CPM functions by identifying the set of activities that control an entire project. Such set comprises of a select few activities among the myriad involved in almost all major infrastructure or construction projects. With the help of CPM, users are able to visually refer to and infer from applied mathematical techniques on planning, analysis, and constant monitoring of all activities in a project. While some activities are linked in series, there are many activities that can simultaneously be running in parallel independent of each other. This form of scheduling also helps in reallocation of resources after rescheduling those activities that are not considered critical to the project. Activity-based scheduling is presented as a bar chart known as Gantt chart in engineering parlance.

Figure 9.

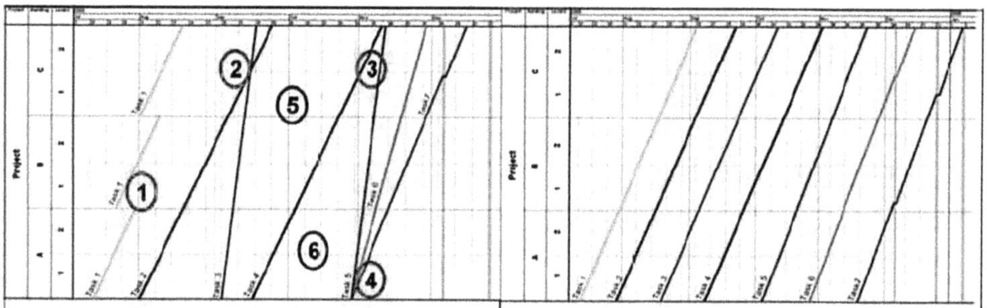

1. Same activity in several locations at the same time	- Continuing tasks of equal duration
2. Crossing of activities	- Standard production rate
3. No time and space buffer	- Activities have time and space buffers
4. Several activities start at the same day	- Slacks at start- and end time
5. and 6. inefficient use of time	- Work divided equally over construction time

Location-Based Scheduling

While Critical Path Method and, in general, activity-based scheduling methods have gained widespread acceptance in projects where there are complex activities each unconnected with the other until reaching assembly stage and areas, there projects that require extensive and repeated fabrication activities *in situ* at various locations. This has given rise to location-based scheduling. This form of scheduling has been a subject of research in academic circles and in the professional realm. However, location-based scheduling has been far-surpassed by activity-based planning and scheduling. As such, Information Technology has not been utilized to promote the development of software tools that can support such form of scheduling. Location-based scheduling is known by various names, viz., Construction Planning Technique, Disturbance Scheduling, Horizontal and Vertical Logic Scheduling, Line-of-Balance, and Time-Location Matrix Model.

Among the above methods, Line-of-Balance offers a visual approach to help planners define and monitor flow of project. This technique uses lines to define the assorted activities that can take place and be acted upon by site personnel at predefined locations. Called Location Breakdown Structure (LBS), is analogous to WBS when considering quantities for building project. The following Figure 9 demonstrates several types of deviations that can be recognized using Line-of-Balance. Advocates of LBS assert that deviations can be identified quickly and remedial action taken to modify linked schedules more conveniently than activity-based scheduling methods, e.g., CPM.

Figure 10. Line-of-Balance diagram illustrating control and monitoring of construction activity.

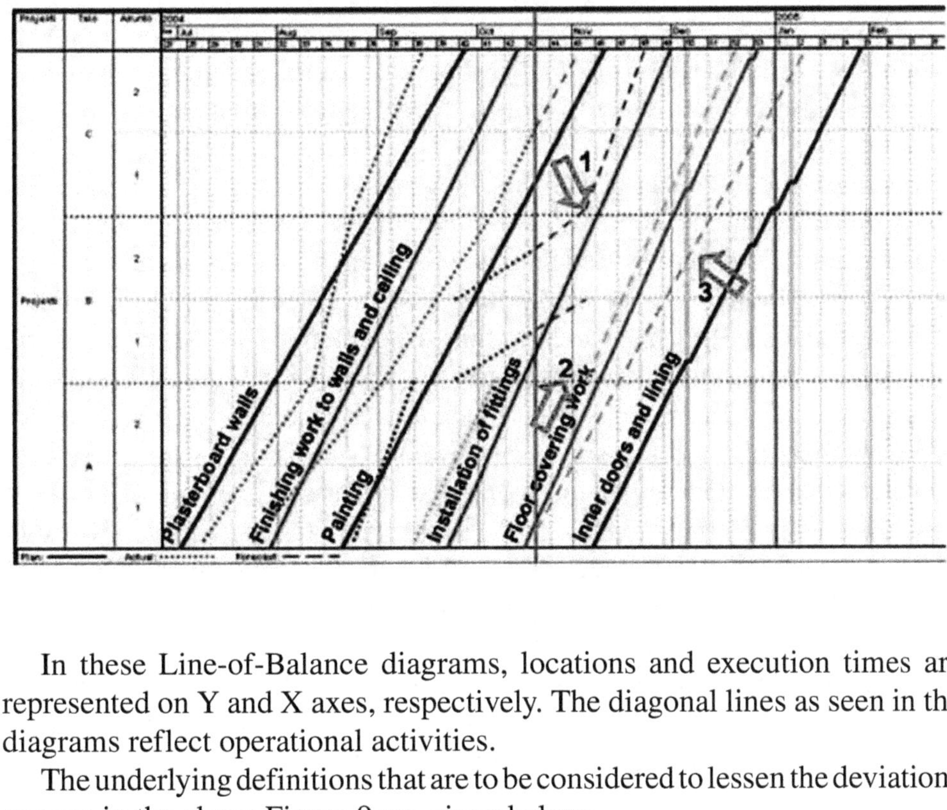

In these Line-of-Balance diagrams, locations and execution times are represented on Y and X axes, respectively. The diagonal lines as seen in the diagrams reflect operational activities.

The underlying definitions that are to be considered to lessen the deviations as seen in the above Figure 9 are given below:

- **Synchronization:** Where planners aim to reach similar production rates for all activities. A synchronized schedule is easily identified by lines running in parallel depicting a constant spatial-temporal gap in related tasks and activities.

- **Pacing:** This implies activities that are planned to carry on to another location unaffected by any interruption. Monitoring of production is key to scheduling projects using Line-of-Balance. What sets this apart from other monitoring activities is that the former involves collecting data on the quantum of activities completed and resources consumed. In case of deviations detected through process of monitoring, clarifications as to the causes for such are asked for and subsequent corrective measures are initiated. Lines in Line-of-Balance diagrams are reflections of actual production achieved as can be seen in Figure 10. The lines are depiction of task histories. These help users find out the location and time on a particular task conducted. Forecasts can then be made on execution of remaining tasks as can be seen in the used of dashed lines in Figure 10. Forecasts, in turn, are essential for carrying out updates to Line-of-Balance diagrams so that required changes to production strategies can be made on demand in future.

As an example, a Line-of-Balance diagram has been presented in Figure 10. In this diagram, the control and monitoring actions have been segregated and identified, accordingly. Red lines have been used for control-moments. As can be seen, production is being disrupted by activities involving installation of fittings and paintings. To address this issue, planner has added resources to accelerate painting as depicted by arrow marked #1 in Figure 10, and work on installation of fittings has been pushed back to a later date as depicted by arrow marked #2 in Figure 10. Slowed down is floor covering works as depicted by arrow marked #3 in Figure 10 to insert sufficient time and space between floor covering work and activities related to installation of fittings.

Using x and y axes, project time and locations, respectively, have been plotted. The lines marked 1 through 3 show control actions that bring about a positive shift in rate of production, splitting and rescheduling of a task, and reduction in production rate, respectively. As a popular scheduling method, Line-of-Balance has gained traction among certain sections of construction industry for the visual acuity and intuitiveness that it provides.

PRODUCTION SIMULATION

Analysis of spatial arrangement and related details of projects are critical in construction projects. While 2D drawings in the form of CPM schedules or Line-of-Balance diagrams are effective, these do have limitations insofar as providing appropriate spatial insights in scheduling processes are concerned. It is this gap that 4D computer-aided design drawing models can address quickly and effectively. Constituting 3D models and CPM schedules, 4D CAD models offer users clarity and focus on the various schedules that can make up a project. Further, 4D CAD models can speed up communication among all participants and stakeholders of a project by allowing each to analyze and carry out simulation specific to respective requirements. What-if scenarios can be constructed using 4D CAD models. However, the most decisive factor in adopting 4D CAD models remains the feature to depict design decision visually, thereby, improving decision making processes as early as in design stages.

It is not unnatural for construction workers to concentrate on their own activities and remain unaffected or unaware of other activities related to project. As a consequence, this result in activities performed out of synchronization which, in turn, require repeated activities to be performed unnecessarily. Not only does visualization encourage workers to be more productive, coordination among various labour gangs are improved. For sharing schedules at macro level, 4D CAD is preferred to communicate using 3D CAD models that are aligned with activity-based scheduling approaches.

ANALYSIS OF CAD

To analyze the efficacy of the two construction planning methods discussed, these were tested in simulations conducted using 4D CAD software. While temporal and spatial issues are undeniably considered extremely significant in any construction project, these are difficult to detect in CPM. On the other hand, 4D modelling method significantly aids in capturing the relevant parameters. It is often suggested that parallel visualization is effective at visually reconciling differences that exist between the two planning methods. In the considered view of many construction professionals, using 4D models helped discern and clearly comprehend assorted processes within a construction project.

FROM 4D TO 5D AND BEYOND

The factor of cost is one that dominates all others insofar as any project is concerned. Indeed, the question of money is foremost for any organization to subsist and survive. Without having a clear picture on costs at any given stage of the project, construction companies will always lack a competitive edge. The feature of integrating cost data has given rise to what is now known as 5D CAD modelling. While many construction companies compute quantity take-offs from paper drawings or even stand-alone digitized versions of the same, 5D modelling offers unrivalled edge and superior processing at significantly reducing computing times.

Table 3. Range of software used by construction professionals

Software Name	Description
MXROAD	This is one of the most advanced software currently available in the market. Offering a sophisticated modelling suite, its users find it intuitive and profitable to work on any project that involves roads.
HDM	Highway Data Management software offers a simple interface that allows users to quickly carry out complex activities on assorted road traffic parameters.
AutoCAD Civil 3D	With a pedigree dating back to the earliest versions of computer-aided design software that engineers have been using since the advent of modern computing technology, this software offers a wide-range of features that can not only aid users in road projects alone, but also helps in land development and water projects as well.
Autoplotter	Used in various diverse construction projects and related activities, the specialization of this software lies specifically in drawing topological maps, 3D contours, cadastral maps, and general surveys related to road construction, mining, etc.
Road Estimator	Road Estimator CAD software is targetted specifically at road constructions.

While many would view process of cost estimation as simply multiplying model object quantities with respective costs, such an approach would be ineffectual and misleading. Individual object models have components that need to be connected to an estimation hierarchy with clear detailing of costs and quantities at basic component level and design stage.

OTHER IMPORTANT SOFTWARE FOR CONSTRUCTION DESIGN-CONSTRUCTION

Road Design and Transportation Software

Information Technology has been elevated to a position of great significance insofar as road design and transportation infrastructure development is concerned. Civil engineers have been using design software that have been developed specifically aimed at carrying out surveying, construction, and maintenance projects. Many of these software are capable of carrying out automatic geometrical computations involving plans and 3D models of roads taking into account terrain parameters. Such functionalities and features have been ploughed in over the years by software manufacturers based on feedback received from users. A table showing various software has been listed in Table 3.

Table 4. Comparison of salient features between WaterCAD and WaterGEMS

WaterCAD	WaterGEMS
Ownership cost of this software is cheaper.	This is comparably more expensive.
The software cannot be run inside ArcGIS.	This can be run inside ArcGIS.
Ability to export to Google Earth is limited.	Data can be exported to Google Earth freely.
It cannot extract elevation data from elevation models.	Extraction of elevation data from digital elevation models is possible.
The software lacks Darwin tools.	Darwin tools (Calibrator, Designer and Scheduler) are part of this software.

Hydraulic Engineering Software

Hydraulics, a branch of civil engineering, deals with fluid dynamics. Projects that involve hydraulic engineering can range from water supply network, pipeline works, drains and conduits to bridges, levees, storm sewers and canals. Advancements made in Information Technology have now made it possible for engineers and researchers to solve numerical problems related to hydraulics in civil construction projects, that would otherwise have been impossible to do within limited time frames.

That water supports life on earth and sustains civilization is undeniable. Equally so is the engineering behind the distribution systems that are required to channel water for human habitats and societies to survive. Before computers began to take on complex numerical activities, slide rules and handheld calculators were mostly used by hydraulic engineers to work on projects that involved water distribution systems and, consequently, fluid dynamics. The extent of the complexity that could be supported, within constraints of time and budget, was, therefore, quite limited in comparison to what is possible today. Modern computing technology has made it possible for experimenting wih 3D models and producing solutions quickly, cheaply and efficiently. There are several such software in the market in use today, e.g., WaterCAD, WaterGEMS, etc.

Table 5. Table showing popular Hydraulic Engineering Software

Software Name	Description
FlowMaster	Mainly used for designing pipes, weirs, channels and inlets, engineers find this software useful to carry out a range of associated computations using this software.
StormCAD	Construction of storm sewers, run-offs, inlets and piping networks can be complex. To make the construction process efficient and cheaper, engineers use this software for modelling and complex calculations specifically targetted at such applications involving storm sewer systems.
CulvertMaster	Used in designing solutions involving culvert hydraulic engineering, this software helps in evaluation and comparison of culverts ranging from the mildly complex to intricate multi-barrel culvert systems.
EPANET	Used for performing simulation on fluid behaviour in pressurized network of pipes, the software is often recommended by engineers for its features to help improve distribution of water.
H2OMAP	The software offers a host of spatial analysis tools, mapping and modelling features for speeding up construction planning and management insofar as hydraulic engineering is concerned.
H2ONET	Users of this software find its powerful features of modelling, analysis and design easy to work in projects involving fluid and fire flow.
SewerCAD	As the name suggests, the software is tailored specifically for projects involving construction and management of sewerage systems.

WaterCAD

Said to be hydrologic software for water distribution system modelling, it is used to determine a range of parameters in respect of pipes, pumps, fire protection systems, etc. Using this software, users can produce various types of reports while also being able to exchange data with other software such as AutoCAD and ArcInfo. The ability to exchange data and information can significantly reduce human resources and time required to make appropriate design-related decisions to build more efficient water distribution networks.

WaterGEMS

Similar to WaterCAD, WaterGEMS allows users to exchange data conveniently with other sophisticated CAD tools such as AutoCAD, MicroStation, etc. Management and modeling of water distribution networks can be done with considerable ease and pace with data that can be shared with Google Earth as well. WaterGEMS is used by engineers to carry out management of many complex and diverse hydraulic engineering activities that would not have otherwise been possible as efficiently in any other tool. A brief comparison of features available in WaterCAD and WaterGEMS software is provided in Table 4.

Table 6. Table showing comparison of salient features betwen SAP2000 and STAAD.

SAP2000	STAAD Program
Created by Computer and Structure, Inc.(CSI).	Offered by Bentley Systems.
Popular in Latin American countries, and in Portugal, Italy and Spain.	Users in Asia and United Kingdom prefer this over any other software.
Its forte is helping in analyses of dynamic and seismic parameters.	Somewhat lacking in dynamic and seismic analyses.
SAP2000 licenses are more expensive.	STAAD licenses are comparatively cheaper.
Supports modelling and analyses involving P-delta, and large displacements.	Does not support p-delta
Not the preferred choice in construction projects involving steel.	STAAD is the preferred choice for construction projects involving steel structures.

In addition to Table 4, a select little other hydraulic engineering software are listed in Table 5 with short description of features.

Structural Engineering Software

As far as critical branches of civil engineering go, structural engineering ranks among the very top. Civil engineers have, over the ages, worked on refining and improving techniques to build large and complex works. Civilizations have largely existed and thrived as a result of such civil construction works that require accuracy and demand resilience and careful eye to detail in designing. The advent of modern computers and rapid strides made in Information Technology has made it possible to analyze designs that would not have otherwise been easily possible by hand or manual computation methods. In turn, this meant that engineers are now able to explore a wide range of approaches to a particular engineering problem and conduct simulations cleanly and efficiently. Structural software enables users to digitally create structural components without involving physical resources to be put in right at the start. Calculations and analysis involving columns, beams, joists, and many more can easily be handled by structural software. There are several structural engineering software that help accomplish complex tasks required of civil engineers today. Among these software, STAAD and SAP2000 rank as one of the best available in the market today.

Table 7. A Brief Discussion on Features of Engineering Software used by Civil Engineers

Software Name	Description
SAFE	Based on finite element method, this software is often used in cases involving analyses of reinforced concrete, post-tensioned slabs, beams, foundations, etc.
RISA 3D	As a 3D modelling software and design analysis tool, engineers often turn to this software in cases of projects that involve common structural materials, viz., aluminium, concrete, steel and wood.
ETABS	The software achieved claim to fame after it was used by engineers to create mathematical models of Burj Khalifa, currently the world's tallest structure.
ANSYS	As a finite element analysis software, it has powerful capabilities that offers assorted range of static and dynamic analyses features.
GT STRUDL	Primarily used in complex infrastructural projects, this software program is used extensively to design and test models involving conventional and nuclear power plants.
LUSAS	This software is marketed as a tool that can help engineers carry out design and analysis involving complex seismic calculations required in projects ranging from offshore constructions to bridges and tunnels.
Tekla	As a versatile 3D modelling tool, designers and engineers bank on this software to create designs for fabrication and erection of complex steel structures.
STRAP	As an analysis and design tool that can support a number of continuum and skeletal structures, e.g., bridges, buildings, towers, etc., its users claim that it is extremely powerful and feature-rich for the purpose for which it has been developed.
Struds	Another structural analysis and design tool that is used in cases where complex calculations are required for construction of beams, columns, shear walls, slabs, foundations, etc.walls and foundations.
MicroStation	Among all the software discussed in this table, MicroStation ranks as the world's top 2D and 3D CAD modelling software that can be used in almost all complex construction projects.

STAAD Program

As far as structural design software tools go, STAAD (Structural Analysis and Design program) has found a firm place in the market for design and construction of large industrial projects encompassing buildings, plants, factories, dams, culverts, embedded structures and many more. One the more unique technical features of STAAD is that it has support for design codes involving concrete, steel and timber. In addition, it also allows users to use several complex analysis tools and approaches ranging from 1st order static analysis through to 2nd order P-delta, geometric non-linear and buckling analyses. It also allows to analyze and engage in model extraction to time-history, as well as spectrum analysis making it one of the more versatile structural engineering software product available in the market today.

SAP2000

Popular structural analysis software is the SAP2000 (Structural Analysis Program). Capable of performing quick equation solving problems, force and displacement loading, Eigen and Ritz model analysis, skewed geometry analyses, etc., the software is considered to be one of its kind available for civil engineers looking for a range of complex features in the market. Primarily used to design buildings, communication towers, dams and any such like complex engineering structures, SAP2000 is preferred by engineers to cut down on costs and time. It also ranks as one of the top user-friendly engineering software available in the market today.

A comparison of two softwares is presented in Table 6.

Some of the other common structural engineering software that are available in the market are tabulated, and their respective features discussed, in Table 7.

Geotechnical Engineering Software

Construction engineering involves soil and water. It is this aspect that Geotechnical engineering deals in. Engineers study geotechnical engineering techniques to work on projects involving earthwork structures. Modelling and analyses tools are widely required in this field to ascertain seismic stability for building structures that need to have solid foundation. This type of engineering also requires careful deliberation and calculations insofar as disposal of waste products by burial is concerned. It is not unnatural to find, therefore, that geotechnical studies concern designs that help in creating soil stabilization systems that are, in turn, integral for the safety and security of life and property. As in the other branches of civil engineering discussed above, handheld calculators were used before complex software tools began to be developed. Today engineers use Information Technology to conduct complex numerical computations that range from the simple to the complex and sophisticated finite- and distinct-element codes. Among the more prominent geotechnical engineering software available in the market today, GeoStudio, FLAC and PLAXIS rank among the best that engineers can use for carrying out design and development work.

Table 8. A Brief Comparison of Features of PLAXIS, FLAC and GeoStudio.

PLAXIS	FLAC	Geo Studio
Developed and sold by PRAXIS BV.	First introduced to the market by ITASCA C.G. in 1986.	Designed and developed by GEO-SLOPE International Ltd.
Simple to use.	Has several features lending to its complexity.	Easy to use.
Results produced are generally acceptable.	Results produced are considered to be good.	Results are generally of acceptable standards.
Friendly user interface couple with steady computation abilities.	Offers basic level features and functionalities making it popular with many users.	Single umbrella software with multiple tools catering to individual requirements.
Targetted for dynamic analyses	Suitable for dynamic analyses	Appropriate for static analyses
Used primarily for embankment, slopes, and tunnels.	Best for tunnels.	Considered best for use in projects involving dams, embankments and slopes.
Expensive to own.	Freely available online.	Expensive to own.

PLAXIS

PLAXIS software was first released as a 2D geotechnical tool targetted to help engineers analyze deformation and stability in case of engineering projects involving soil. Projects ranging from dikes, tunnels, slopes and embankments are appropriate areas where PRAXIS software can be used. PRAXIS is often preferred for its simple and intuitive interface. A 3D version of the software was released in 2010 and quickly came to be known as flexible 3D software that are targetted at users not proficient in numerical analysis. The software offers robust geotechnical workflows, an easy to use graphical interface, and a borehole wizard for soil modelling and 3D simulation.

FLAC

FLAC (Fast Lagrangian Analysis of Continua) is a powerful 3D software that was first introduced as a 2D design tool in 1986 by ITASCA, C.G. Based on finite-different meothd, as a general analysis and design software, it utilizes recent advancements made in multi-stage micromechanics. This software offers 13 built-in models and various plotting and statistical distribution features. It also offers built-in scripting language that can customize and automate almost any of the features that the software offers.

GeoStudio

GeoStudio software offers a suite of tools catering to geotechnical, environmental, civil and mining projects. The software was developed and is sold by Geoslop International Ltd., of Calgary, Alberta, Canada. Engineers use it mostly for finite slope analysis, slope stability computations, analysis of seepage, etc. It is actually an amalgam of eight distinct software tools, viz., SLOPE/W for dealing in slope stability, SIGMA/W for dealing in stress-deformation, QUAKE/W for computations related to dynamic earthquake simulations, TEMP/W for geothermal calculations, CTRAN/W for transport of contaminants, AIR/W for calculations related to air flows, and VADOSE/W for analyses related to soil covers and vadose zone. These tools are integrated and together form a synchronized mesh allowing users to port results from one software to another. As a whole, the software is intuitive and offers a moderate learning curve that most find easy to adapt and master. The Table 8 lists salient features and differences among PLAXIS, FLAC and GeoStudio.

REFERENCES

Akbas, R. (2004). *Geometry-based modeling and simulation of construction processes* (PhD Thesis). Department of Civil and Environmental Engineering, Stanford University.

Akinci, B., Fischer, M., & Kunz, J. (2002). Automated Generation of Work Spaces Required by Construction Activities. *Journal of Construction Engineering and Management, 128*(4), 10. doi:10.1061/(ASCE)0733-9364(2002)128:4(306)

Amor, R., & Faraj, I. (2001). Misconceptions about integrated project databases. *ITcon Electronic Journal of Information Technology in Construction, 6*, 57–68.

Andersson, N., & Suber, K. (2006). *Konfigureringsprocessen vid industriellt byggande Förutsättningar för ett framgångsrikt industriellt byggande med fokus på kundanpassning av hus. INDEK, Royal Institute of Technology.* KTH.

Graphisoft, Inc. (2006). Virtual Construction - Graphisoft ArchiCAD, Constructor, Estimator *Presenter*. www.graphisoft.com

Chapter 3
The Role of Software in Construction Management

ABSTRACT

Civil construction projects not only need to be executed on time; these also require that budgetary overruns are not allowed to take place needlessly. The traditional constraints involving time, quality, and money must always be paid close attention for a project to be considered commercial and engineering success. In this chapter, the authors discuss the role of project management software and the various commercial options available in the software market for entrepreneurs, engineers, and project planners to explore. They also discuss the need for activity codes and project scheduling types and the significance of these in civil construction engineering. The importance of planning and scheduling cannot be overestimated in a world where competition is high and civil construction companies often find themselves on razor's edge to stay afloat and remain profitable.

The construction industry has been subjected to complex changes over the past decades. Industrialization, mechanization and infusion of Information Technology has transformed the way civil construction companies do business today. Global competition has long prodded construction companies to adopt sophisticated operational and business practices. Given the complexity of civil construction projects, no longer is the term "construction" associated just with human resources, material resources and machinery. The term today has grown synonymous with initial conception, planning, materials

DOI: 10.4018/978-1-7998-5291-9.ch003

and resource management, and execution. It is well understood that for a civil construction project to succeed, planning processes must take into account market surveys, negotiations with vendors, contract management, choosing the method of construction, arranging for finances and cash flow at critical phases of construction, and all related sub-activities covering supply chain management.

The nature and level of complexity involved in executing civil construction engineering projects are far too complicated to solve using traditional methods of yesteryears. The only route for civil construction engineering companies to thrive and flourish in the modern competitive world, is to adopt newer Information Technology tools and techniques as these become available. Planning and scheduling are the most important features of any project planning. Sequencing of activities and monitoring progress as project execution continues is another vital aspect in project planning.

CONSTRUCTION MANAGEMENT – A BRIEF HISTORY

Construction Management as we know today has evolved from a merger of associated concepts, practices and ideas stemming from construction, engineering and defence. It is said that construction management is the child of Henry Gantt, who is the USA first developed a bar chart as a tool for managing and monitoring construction activities. His work gave rise to the important work breakdown structure (WBS) among many other modern construction management techniques and tools known today. Progressively, a novel approach known as the "milestone method" was developed to gauge accurately progress of a project. The individual bars on a progress chart were referred to as milestones which gave rise to the "milestone method". To assist in handling serial methods of production as in a production line, a "line of balance" technique was developed to handle repetitive tasks.

Sometime in 1957, a critical path scheduling (CPS) technique (Bennett, Laurence, 1977) was developed that could be supported by computers and latest advancements that were then being made in the field of Information Technology. Later, the CPS was modified to form critical path method (CPM). At around the same time, two important mathematical models were developed for project scheduling:

1. Booz-Allen & Hamilton developed the Program Evaluation and Review Technique (PERT) in association with the Lockheed Corporation for

conducting work on the US Navy's Polaris nuclear submarine missile programme.
2. The DuPont Corporation and the Remington Rand Corporation jointly set about to develop the famed critical path method (CPM) for management of maintenance at plans.

It did not take long for these mathematical approaches (Kumar, 1998) to be adopted by the private sector in order to boost manufacturing processes and grow more efficient in rapidly developing markets and economies. These two mathematical approaches further gave rise to precedence networks that are considered as specialized techniques to be applied in repetitive or serial production environments. Graphical Evaluation Review Techniques (GERT) was also developed at around this period that catered to more than a single uncertainty in a given system.

The progression of Information Technology in terms of hardware and software (Kumar, 1990) resulted in production of more powerful and cheaper computers. Projects that were considered too big for traditional computing methods to handle now began to be handled with relative ease using Information Technology. Information Technology now makes it possible to plan and monitor progress of not only resources alone, but also of time and scheduling of project activities (Lewis, 2002) along with individual sub-activities on a micro scale. To serve the interest of construction management and construction industry, the Project Management Institute (PMI) was created in 1969 in the USA. In 1981, the PMI began work on and soon published "A Guide to the Project Management Body of Knowledge" (PMBOK). This guide was written to address all guidelines and standards of practice which are followed in the construction industry and its management. Founded in 1967, a European organization similar to PMI took shape. The International Project Management Association (IPMA) worked hand in hand with PMI to shape modern global project management standards that are adopted worldwide.

WHAT IS A PROJECT MANAGEMENT SYSTEM?

The primary objective of a project management software is to provide a stable platform for organizing, strategizing and forming estimations that can be as accurate as possible. A sophisticated project management software can have as many features as possible for planners and engineers to take into account resource breakdown structures, resource availabilities, rates to name a few.

Depending on nature and complexity of project management software can be any of the following types that are considered major ones in the industry:

1. With an intuitive graphical user interface, desktop-based project management software programs give an edge to organizations and users when it comes to responsiveness.
2. Web-based project management software is more or less akin to a system of environment wherein users need only to use the software using compatible browser installed in any computer. No software installation is necessary outside the installation of browser. While this is an advantage in cases where users may lack proper computer hardware to support a full-featured desktop application, such kind of software usually are not quite as responsive nor can such web-based software be used in case network connections are absent.
3. Often project management systems can be basic and useful for individuals in case of handling simple projects for domestic use.

PROJECT MANAGEMENT

For a project to succeed, a proper planning is essential. In turn, for planning to be relevant, the factors of time, cost, and project scope must be considered. A well-thought out plan should include measures that are to be initiated in case of risks to mitigate any negative effects. In a construction project, project planning encompasses assessing and understanding requirements, formulation of WBS, defining activities and relating subs-activities, sequencing of activities, estimating duration, etc.

Planning is an amalgam that constitutes assorted knowledge and application areas, descriptions, and sub-activities or related processes. The following illustration as in Figure 1 is produced below to offer a clearer understanding of various aspects of planning.

Figure 1. Illustration of Knowledge Areas in Relation with Processes

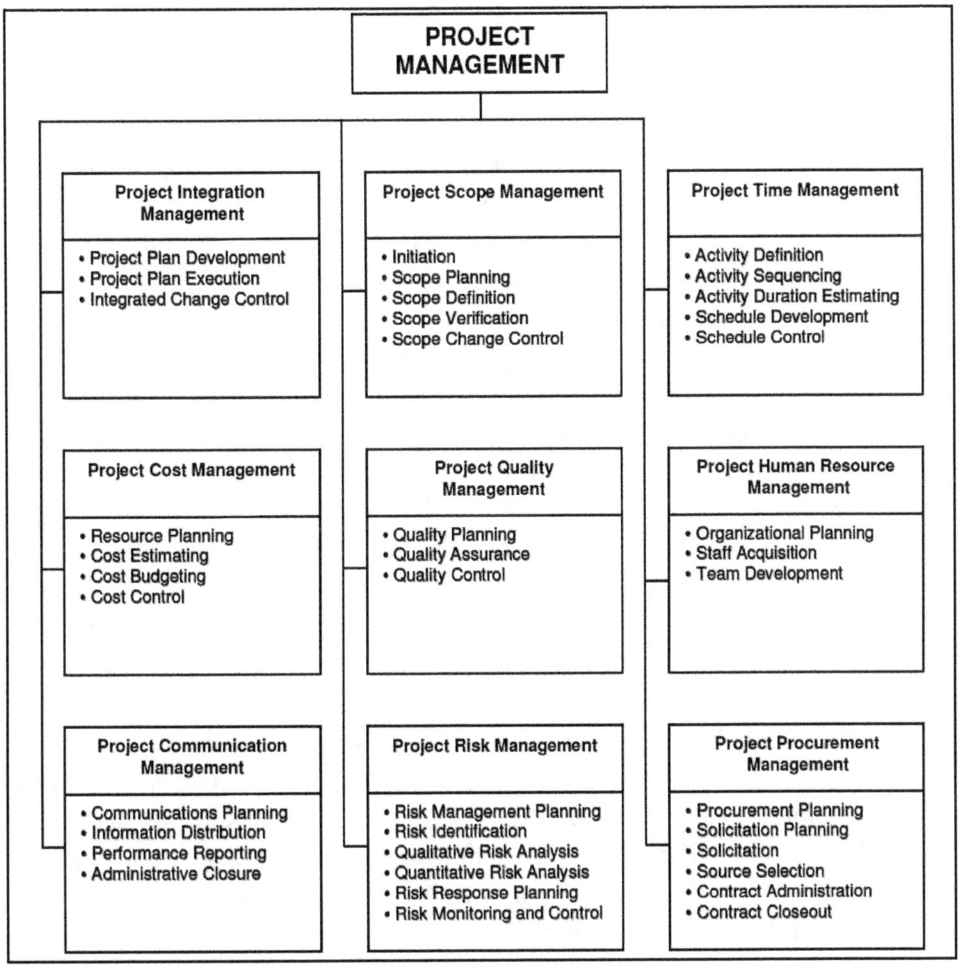

The salient points are defined in the manner as follows below:

- In the Project Integration Management phase, development and execution of project plan takes place.
- The Project Scope Management stage defines and describes activities involved and required to complete a project successfully. Planning, initiation, definition, verification, and change control of scopes are taken care of in this stage.

- Defining activities, sequencing of activities, estimation, scheduling and control are described and contained in Project Time Management phase.
- Resource planning, budgeting, and estimation of costs are part of Project Cost Management phase.
- The Project Quality Management phase is considered by many to be one of the more important ones among those discussed here. In this phase, activities are included to ascertain the level of quality achieved, methods to enforce quality control, and related process to assure levels of quality as planned prior to commencement of the project.
- Human resources is considered as the life blood in any organization. It is, therefore, only but natural that project planning must take into account stages of planning involving staff, training, development of teams and organization of such into the framework of a given planning stage. The Project Human Resource Management is the phase wherein all factors related to human resources planning are considered and accounted for.
- In the Project Communication Management phase, planning of communication, reporting, administrative actions, and distribution of information are included.
- Project Risk Management is concerned with identification, analyses and carrying out coordinated responses to risks. It covers an array of risk-related activities that include, but is not limited to, qualitative risk analysis, quantitative risk analysis, risk response planning, risk monitoring, risk identification and risk management planning.
- Procurement planning, planning of solicitation, selection of sources, contract administration and allied contract related activities are taken care of in the Project Procurement Management phase.

Earned Value Management

For success of a project to be measured, some form of a method needs to be adopted. The Earned Value Management method is widely considered as being one of the more acceptable methods in industry circles. In order to compute earned value, it is necessary to consider the following parameters:

- Planned Value (PV)
- Earned Value (EV)
- Actual Cost (AC)

As the name implies, Planned Value (PV) represents the budgeted cost for an activity or an entire project for a set time period. Budgets and estimates are made during Planning Process. The cost of direct and indirect costs related to project are represented by Actual Cost. The value of work completed as on a given date is classified as the Earned Value.

The constituents of Earned Value Management are depicted in the Figure 2.

Figure 2. Earned Value Management

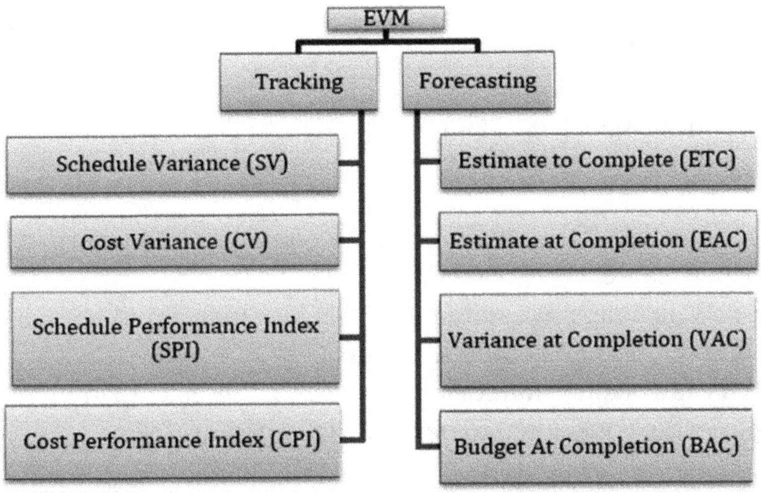

Tracking

For any project to succeed, planning has to take precedence. However, in order to determine whether project progress is on track, some form of tracking methodology has to be put in place. The purpose of tracking is to stay as closely aligned with plans setup as project execution continues. During the process of tracking, project progress is analysed in terms of resources used and costs involved. Using the Earned Value Management method, analyses are conducted to ascertain whether there have been cost overruns or resource over-utilizations. Under-utilizations, if any, are also recorded and reported. In case of deviations in schedules, costs or resources, these can be identified and control measures initiated. The following formulae are used to determine variances:

Schedule Variance (SV) = Earned Value (EV) – Planned Value (PV)

Cost Variance (CV) = Earned Value (EV) – Actual Cost (AC)

Schedule Performance Index (SPI) = Earned Value (EV) / Planned Value (PV)

Cost Performance Index (CPI) = Earned Value (EV) / Actual Cost (AC)

Schedule Variance Cases

Schedule Variance can be behind schedule, as per schedule, or ahead of schedule. Formula in each of the three cases is given below:

Case 1: If SV = 0, SPI is considered to be equal to 1 and project progress is as per schedule.
Case 2: If SV is a negative number, SPI is less than 1 and project progress is considered to be behind schedule.
Case 3: If SV is a positive number, SPI is said to be greater than 1 and progress of project is considered to be ahead of schedule.

Cost Variance Cases

Cost Variance cases can be as per budget, may have budget overruns, or remain below budget. Formula in each of the three cases is given below:

Case 1: If CV = 0, CPI is considered to be equal to 1 and project costs are said to be as per budget.
Case 2: If CV is a negative number, CPI is less than 1 and project costs are said to be over budget.
Case 3: If CV is a positive number, CPI is said to be greater than 1 and project costs are said to be under budget.

Forcasting of the Project

Complementing the process of tracking is the feature for forecasting that is also supported by Earned Value Management (EVM). Forecasting helps in determining cost estimates and also costs that might be required till project completion. The process of forecasting is also useful to estimate performance of a given civil construction project.

The following formulae are used in forecasting:

Case 1: If variance is not expected to carry on.

Estimate to Complete (ETC) = Budget at Completion (BAC) – Earned Value (EV)

Estimate at Completion (EAC) = Actual Cost (AC) + Estimate to Complete (ETC)

Estimate at Completion (EAC) = Actual Cost (AC)

+ [Budget at Completion (BAC) – Earned Value (EV)]

Variance at Completion (VAC) = Budget at Completion (BAC)

– Estimate at Completion (EAC)

Case 2: If variance is expected to carry on affecting only cost.

Estimate to Complete (ETC) = [Budget at Completion (BAC)

– Earned Value (EV)] – Cost Performance Index (CPI)

Estimate at Completion (EAC) = Actual Cost (AC) + Estimate to Complete (ETC)

Variance at Completion (VAC) = Budget at Completion (BAC)

– Estimate at Completion (EAC)

Case 3: If variance is expected to carry on affecting only cost and schedule.

Estimate to Complete (ETC) = [Budget at Completion (BAC) – Earned Value (EV)]

/ [Cost Performance Index (CPI) * Schedule Performance Index (SPI)]

Estimate at Completion (EAC) = Actual Cost (AC) + Estimate to Complete (ETC)

Variance at Completion (VAC) = Budget at Completion (BAC)

– Estimate at Completion (EAC)

TRADITIONAL CONSTRAINTS

In case of every project, commitments to time, cost and scope must be met in order for such to be deemed a success. In turn, these are the very factors that are traditionally referred to be many as trip constraints of any project. These constitute the vertices of what is known as Construction Management Triangle wherein each of the vertices represent a constraint. Any change to any one of the constraints will impact the remaining two. Refining of the vertex scope results in quality or performance that is considered to be a fourth constraint. An alternative approach to Construction Management is to consider Time (T), Quality (Q), and Money (M) as three constraints. The TQM acronym is often considered to be in synch with Total Quality Management (shown in the following figure 3) for ensuring quality in both manufacturing and business practices.

It, therefore, follows that if more human resources are hired a project job may be finished quicker; but it will also raise costs. Alternatively, if the objective is to keep costs and human resources constant, a task needs to be completed in shorter amount of time.

Figure 3. Vertices of Total Quality Management triangle

The Argument for Information Technology

Today, the demand for complex civil engineering projects is on the rise. As human civilization moves forward, civil construction engineers are pushed to create buildings and structures to improve quality of life. Civil construction engineers and managers are now required to deal with data and information streaming in from multitude directions. Appropriate and secure handling of data becomes imperative. It has long since been established that Information Technology is the way to go in such case.

THE INFLUENCE OF INFORMATION TECHNOLOGY IN CONSTRUCTION MANAGEMENT

Civil construction engineering projects often involve numerous tasks. Tracking of all the tasks involved is in itself a complicated process. Scheduling and diagramming of such make for an even more challenging prospect. As such, the influence of Information Technology in helping to effectively handle construction management projects becomes extremely relevant and a matter of extreme interest.

Project Planning

Much thought has to be given while preparing for project planning itself. Attention needs to be paid to number of activities involved, use of time units and calendars, inclusion of Gantt and PERT charts, planning for sub-networks and nested networks. However, the most important aspect of any project planning is to create provisions for rescheduling or carrying out automatic updates.

Resources Management

Managing resources involves taking into account number of resources, resource sharing and levelling, scheduling of resources, load resource updates, resolving resource conflicts, assigning and managing multi-project resources, cost estimation and modelling financial analyses.

Tracking / Monitoring

Tracking and monitoring involves conducting analyses based on critical path and sub-networks. It also includes careful consideration to be given to early warning systems, baselines and schedule update, and display of like items.

Report Generation

Sophisticated data analyses, report generations, resource detailing, development of Gantt and PERT charts and diagrams are integral parts to this phase.

Decision Making

Decision making activities that often have What-If analyses, multi-project and cross analyses, and involvement of Information Technology in an extremely significant way are involved at this stage.

PROJECT MANAGEMENT SOFTWARE (PMS)

The increasing complexity and sophistication levels of civil construction and engineering projects have necessitated the use of Information Technology in more ways than could have ever been imagined in earlier times when traditional methods held sway. As an indispensable tool for organizing activities related to complex projects, the use of project management software (PMS) has grown in stature. The construction sector, today, cannot work without a PMS. In fact, without a PM software it becomes almost inconceivable that assorted activities and sub-activities can be carefully orchestrated to produce results effectively and efficiently. In turn, Information Technology can help in developing even more capable PM software based on the inputs of civil engineers and construction professionals interested in improving software for effecting improved project planning and management. A PMS often has the kind of flexibility and advantages that cannot be rivalled by traditional methods. A user can intuitively work on a PMS and generate reports using graphical means saving considerable time and effort.

Most available PMS currently sold in the market offers users graphical user interfaces that enable any user to work freely without requiring any understanding of the underlying complexity. Broadly speaking, there are three different types of PMS that are used by engineers in construction sector. The first of these types is not particularly well-known for exhaustive set of project management features. Mainly used for scheduling, Microsoft Project is an example of this type of PMS. The second of the PMS types is Oracle's Primavera. Able to handle complex portfolio management activities, Primavera can be used for exchanging data and information with enterprise resource planning (ERP) applications. The last type of PMS is one that integrates on-site management with the usual practices followed in the construction sector. The PMS types are discussed in the following sections.

Microsoft Project

Developed and sold by Microsoft, Microsoft Project (MSP) offers several features for project managers to help them create project plans, assign resources to specified tasks, track progress, analyse workloads and manage budgets. The first version of MSP was a very basic affair that, first launched in 1984, worked on MS DOS. At that point, the software was produced by a different company. Taking note of the potential of the software, Microsoft acquired the product soon after before the end of the 20th century. Today, Microsoft Project ranks among the more popular choice of PMS for construction engineers and professionals. However, in spite of its popularity, it has failed to become the top choice the software is still not completely tuned and aligned with unique set of requirements of construction industry.

Microsoft Project is capable of supporting critical path with support for critical chain and event chain methodology as third-party add-ons. The software is also capable of distinguishing different user classes with support for different user access levels to projects, views, and other data available. The software also has visualization feature available for Gantt chart. Custom calendars, filters, fields, tables, and views can be shared among all users.

Approaches to Microsoft Project Software

1. **Desktop** - PMS can be implemented as a desktop application. The advantage of a desktop-based application is the availability of rich graphical features available in the interface. Desktop-based applications can store data locally in files on hard drives, or at a central location over a network.
2. **Web-based** - Microsoft also offers MSP in a web-based variant. Accessible over networks using a web browser, it shares the usual set of advantages and failings of normal web-based applications:
    a. The application can be accessed from any PC without the need to install full-fledged software.
    b. Access control is easier to enforce.
    c. Web applications are by standard multi-user.
    d. Installation and maintenance of software version is greatly simplified for all users.
    e. However, web applications are often found to be slower than their desktop counterparts.
    f. Graphical features are limited than those of desktops.
    g. Web-based applications do not function well in offline modes.
3. **Single user** - Usually project management software are designed to allow users to collaborate in real time or shared files among themselves offline. However, in select few cases, single user project management systems are desirable. This type of software is used in very small companies where project planning involvement is limited only to one or two individuals at the top of managerial hierarchy.
4. **Collaborative** - Web-based tools are popular among users because of their concurrent multi-user support facilities. However, the chief constraint for such web-based software and tools is the availability of

network and/or network stability. In order to mitigate the effects of poor network connections, software developers have come up with a Rich Client application system that works on client-server model. Data and information are synchronized with a central server or data center at periodic intervals.

5. **Integrated -** Integrated PMS usually involves not only project planning and management, but also keeping track of bugs and issues. Each user can have own separate lists of task activities, calendars, messaging functionalities associated with separate projects.
6. **Non-specialised tools -** Although users continue to prefer one PMS over another, there are software tools available in the market that are often used to plan and schedule project activities. These non-specialized software, or methods that are not related to Information Technology in any way, have inherent limitations. For example, regular calendar software can be used to prepare schedules. Far more effective than ordinary calendar software can be spreadsheets. Spreadsheets are excellent tools to work on any complex numerical or data related calculations. However, these are still some way short of the specific requirements that only dedicated project management software offer.

Benefits of MSP

Microsoft's project management software, commonly referred to as MSP, offers a web-based version that has gained a substantial fan-following among project managers and professionals working in civil construction sector. Supporting advanced features and multi-dimensional spatial workspaces available in case of almost all types of construction activities, engineers and construction professionals consider MSP to be a vital tool in projects involving significantly heavy workloads. Microsoft Project is decidedly a step up in project management. Enabling users to view cascading changes even in case of the most minor alteration in any related activity, MSP also allows to change data at any given moment with considerable ease.

The software, i.e., MSP, functions based on the concept of Gantt Chart. This enables it to allow users to discern easily progress of work within schedule. However, it does not support earliest start and finish times – also referred to as EST and EFT, latest start time (LST), or latest finish time (LFT), etc. Nevertheless, it shows float which is essential in case of resource levelling. The function of floats is to level peak resource requirements on any given day.

With the advent of MSP, planning managers and civil construction engineers began to explore more features than that available in basic Gantt Chart. Critical path and critical activities are supported in MSP. So are milestones, assigning zero times on Gantt Chart and control can be established firmly on various tasks for completion within stipulated period. Web-enabled features enhance collaboration and effectively reduce communication gaps between personnel. Along with the above points as discussed, such features make Microsoft Project eminently suited for any form of construction project.

PRIMAVERA (P6) IN CONSTRUCTION INDUSTRY

Primavera P6 from Oracle is considered as a serious and worthy contender to the crown of the most feature-rich project management software available in the market. Primavera enables users to create and form plans concerning tasks, resource allocations, tracking of progress of related activities, managing budgets, etc. The software can be used to create work breakdown schedules, as also schedules for assigning cost to each task in a project. The software is extremely useful where EVM techniques are suitable and CPM scheduling is required.

The software was originally developed and sold in the market way back in 1983 by Primavera Systems, Inc. Offered as one-of-its-kind enterprise project portfolio management system the company went on to be acquired by Oracle two decades later for its enhanced integration features that allowed users to exchange data and information with Oracle's ERP and SAP's ERP environments.

Procedure in Primavera

- **Create EPS (Enterprise Project Structure) - One** of the key initial steps in Primavera, is to first collect all data available that are relevant and pertinent to the project. The relevant steps that are to be followed in Primavera P6 are detailed and described in subsequent section. Completion of structure for the company along with all its branches is necessary. This is known as the Enterprise Project Structure.
- **Organizational breakdown structure (OBS)-** An organizational breakdown structure, also known as the OBS, is created soon after

creation of the EPS. The OBS is essentially a hierarchy of responsibilities and roles for projects and project-related activities in an enterprise.

Figure 4. A Primavera window showing Organizational breakdown structure.

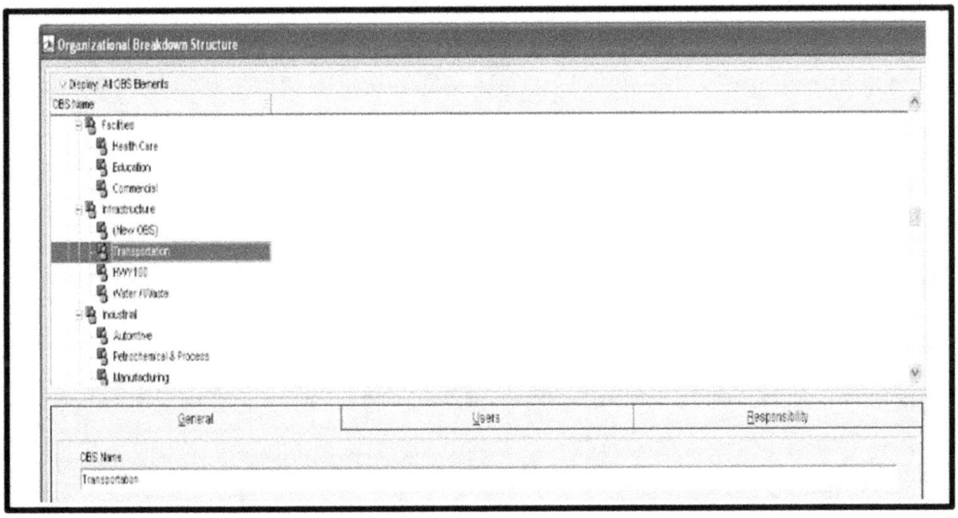

- **Create Project -** A project is a set of assorted activities and data that together form a plan for producing any product or resulting in a service rendered. Under EPS, a project is created under designated sections. A project is assigned planned start and finish dates. It is also assigned to a calendar that is either a global, resource or project one.
- **Define WBS -** Work Breakdown Schedule elements help in demarcating the various deliverables, and assist in creation of reports related to schedules of projects. Significantly, WBS help in ascertaining approximate costs at various level of details. In other words, WBS represents a hierarchy of activities that exist in a project and is a representation of the elements that need to be produced for an entire project to read its conclusion. While every project has its own WBS with associated hierarch structure, the top WBS element is always at the same level and equal to that of its corresponding EPS node. Every WBS element contains details of WBS levels, activities, resources as well as constraints identified in the project.
- **Creating Calendar -** Activities are assigned to respective calendars. The purpose of such is to define work hours that are available in each of

the days in respective calendars. Holidays, working and non-working days are defined in calendars.

Figure 5. Example of Calendar

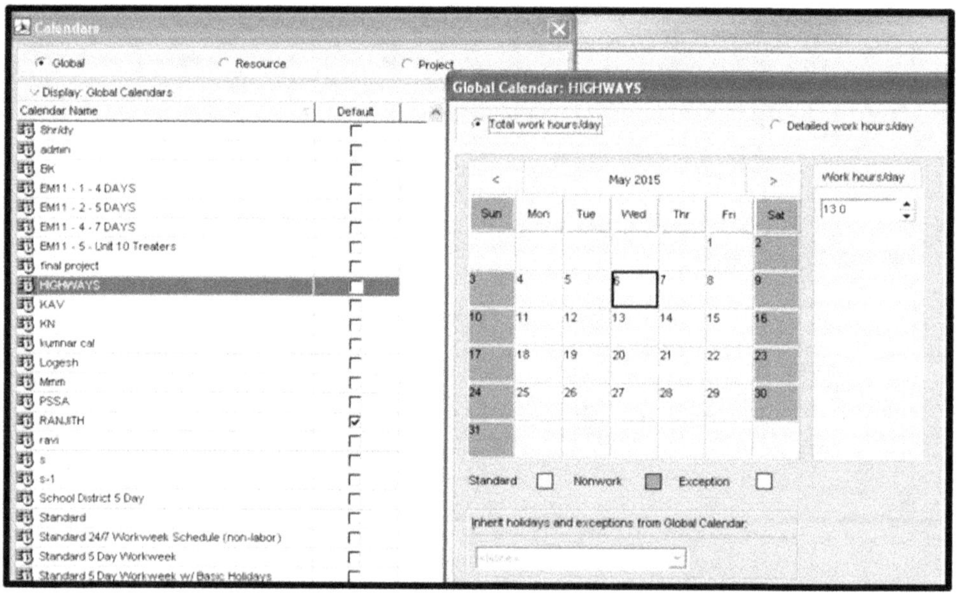

- **Define Activities -** Activities are described as the basic building blocks of a WBS. Activities constitute the lowest through to the top-most level of a WBS. An activity has characteristics encompassing its ID, its name, beginning and end dates, calendars, activity codes, type, constraints, costs, relationship relating to its predecessors and successors, resources assigned, and roles.
- **Durations for every Activity -** Project durations are entered in appropriate fields when starting off with planning of a work. Actual durations, however, can only be fed for the respective project activities only upon completion.
- **Activity dates -** Primavera supports actual start, actual finish, planned start and planned finish activity dates.
- **Activity cost -** All costs related to an activity is summed under Activity Cost.
- **Relating Every Activity and Perform Scheduling -** Networks are formed by inter-connecting scheduled activities. This is usually

accomplished by analysing preceding and succeeding activities and assigning these to form relationships within the scope of overall project. Following is a list of relationships that may exist:

  - Start to Start (SS)
  - Start to Finish (SF)
  - Finish to Start (FS)
  - Finish to Finish (FF)

- **Scheduling** - Scheduling is performed using Critical Path Method (CPM). Activities, durations and relationships are defined using CPM. Using scheduling interconnected activities can be examined to calculate total time needed for project execution. The CPM approach is among the more preferred ones to have been adopted by a major chunk of organizations for schedule management requirements in the construction sector.
  - **Critical Path Method (CPM) Scheduling** - Durations of activities and relationships measured between activities are used in CPM for scheduling dates. This involves two passes on activities within a project. As such, the following points are of interest:
    - The critical path defines the duration of a project and the estimated completion date of project.
    - A delay in any of activity leads to cascading effect in other activities resulting in delay of the project as a whole.
    - Primavera enables users to compute critical path in terms of longest path or total float in the entire project.
  - **Data Date** - The "data date" is the date on which the last edit has been made in schedule before it is handed over an engineer by the project scheduler. This date usually coincides with the last working day of the contract period of the project. Using the schedule tool, a project scheduler can carry out edits to project data date.
- **Scheduling Concepts**
  - **Forward Pass:**
    - Computes early dates of an activity
    - The earliest times an activity can commence and end after its predecessors have been completed, are known as early dates.
    - Computations begin on activities without predecessors.

Early Start + Duration - 1 = Early Finish

- **Backward Pass:**
  - Computes late dates of an activity
  - The latest times an activity can commence and end without causing any delays to the end date of project, are known as late dates.
  - Computations on activities begin without considering successors.

Late Finish - Duration + 1 = Late Start

- **Total Float**
  - This is considered as the amount of time that an activity can slip from its early start without affecting or delaying project.
  - Total float is the difference between activity late date and activity early date.
  - Zero total floats are considered critical.

Late date – Early date = Total Float (TF)

- **Benefits of Scheduling:** It has been observed that CPM has been found effective for scheduling and for meeting project objectives. Followers of this method of scheduling claim that it has a positive influence in promoting logical thinking, helping in evaluating options, improving communications, tracking progress, identifying delays, facilitating recoveries and meeting legal requirements.
- **Scheduling Methods -** A range of scheduling methods are available and used by construction industry. Several of these are listed below:
  - **Bar Chart -**Bar charts are basically schedules implemented in two dimensions. Timelines of projects are represented by the X-axis, while the activities of the project under consideration is shown on the Y-axis. Activities of a project are arranged in the order of earliest start. It is because of this order that almost all bar charts display bar patterns beginning from upper left of chart through to bars that appear at the bottom right of it depicting completion of project. One notable disadvantage of bar charts is the implied sequencing of activities. Put simply, there are no visible or indicative constraints governing one activity to commence without any reference to another related activity upon which the former is dependent.

Figure 6. Window Showing Distribution of Roles

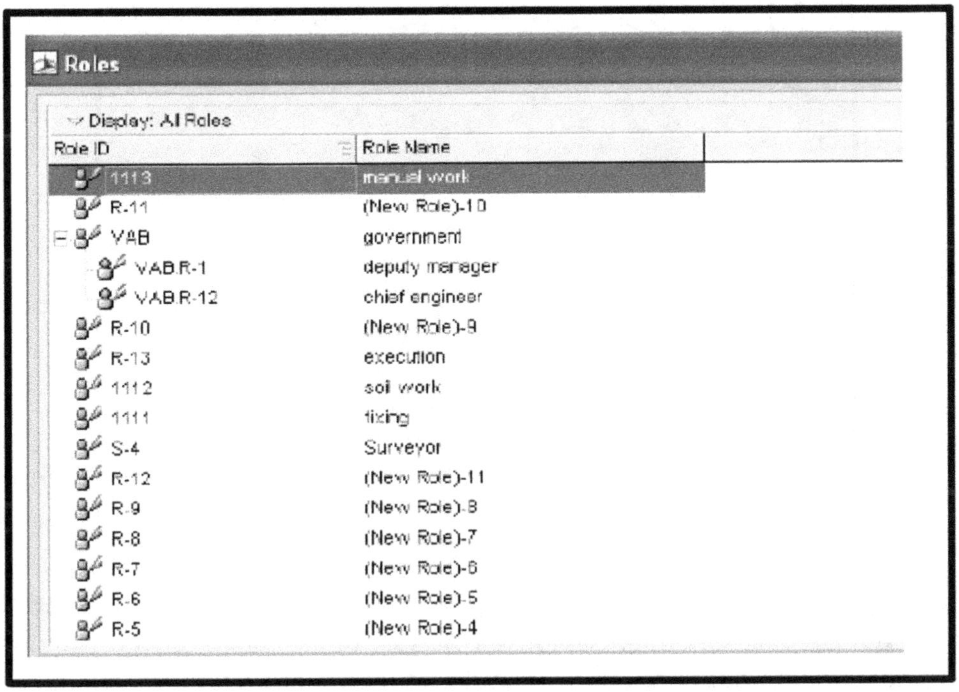

- **Fenced Bar Chart (Time-Scaled) -** Fenced bar charts are a step up from regular bar charts in that the schedule of activities can also be depicted along with sequence. The schedule and sequence of activities are connected using arrows to illustrate schedule and sequence.
- **Arrow Diagram -** Combining sequence and schedule, arrow diagrams are often used to proportionately describe duration of related activities. The arrowheads signify end of an activity whereas node indicates the start of an activity.
- **Precedence Diagram -**In precedence diagram method (PDM), schedules are shown with activity durations in box or block. Relationships between two activities are shown using arrows. Arrowheads are indicative of the direction of schedule involved. Direction of travel is usually from left to right corresponding to start of a project till the end. Activities precede one another as per logical sequence and as per schedule.

Figure 7. A Sample Project Code Window

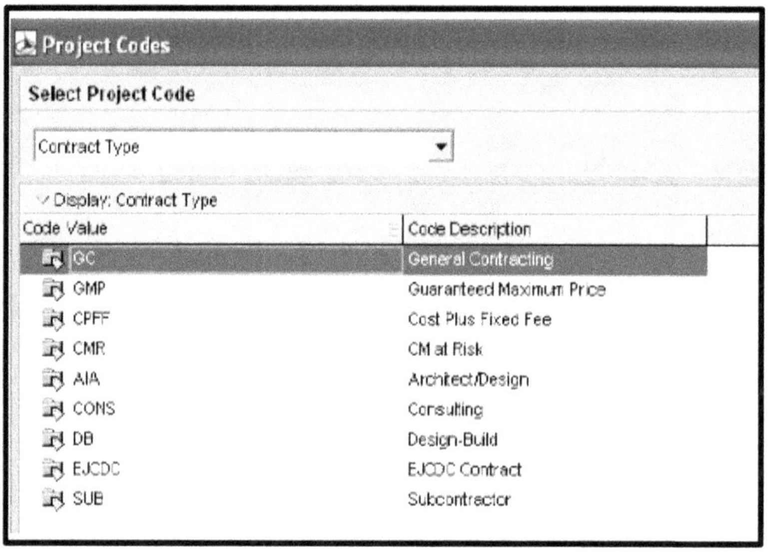

- **Role -**Roles are representatives of job titles and often correspond to specific skills required in order to carry out activities related to project. Examples of roles range from architects, engineers, quality assurance testers, to skilled and semi-skilled workers. The following figure 6 demonstrates how a set of roles can be created and assigned to activities in a project. Hierarchical arrangement for management and assignment are also shown therein. An activity's skills requirements can be analysed based on roles defined.

- **Project Codes -** The purpose of project codes is to relate projects based on common characteristics. In cases where EPS may contain several projects at assorted hierarchical levels, project codes are useful in maintaining the desired arrangement. Project codes come in handy when filtering through information in order to group and display projects that may span across different areas in an enterprise. The following figure 7 demonstrates how project codes can be used to filter, sort and prepare reports on projects arranged hierarchically.

Resource Codes

Resource codes are invaluable means of tracking any number of resource types that an organization may use. Resource codes are used for efficiently tracking, reporting and analysing resource utilizations. The following figure 8 shows resource categorizations using codes.

Figure 8. Window showing Allocation of Resource Codes with Descriptions

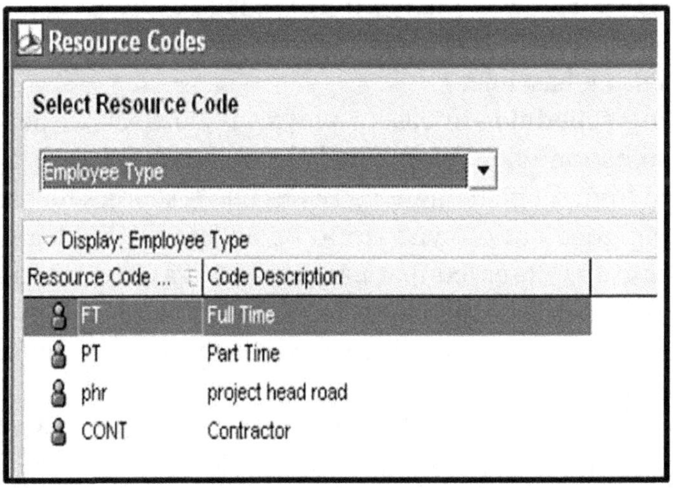

Activity Codes

There can be up to three types of activity codes that help group activities in to logical blocks in accordance to an organization's criteria. Activities that are common for all projects are clubbed under Global Activity Codes. Activity codes that group activities within a certain branch in an enterprise project structure (EPS) are called EPS activity codes. Project activity codes are defined in accordance with certain set features that are identifiable within a project. Activity codes are representative of categories of information that may even include a department or an individual. While proceeding with creation of activity codes, care should be taken to identify categories first and then allocate values against each code. Activity codes are similar to tags with each activity code attached to corresponding task or activity, thus, facilitating changes to be made to the data and tracking. Activity codes can represent several nuggets of information such as divisions and phases of work to be performed, or even connect to locations where activities are to be carried out.

- **Levelling of Resources -**Free float activities are taken into account when computing critical path for a planned schedule. Such activities are usually accorded top priority during allocation of resource to save time. Accordingly, resource levelling and critical activities are also accorded a very high priority.
- **Creating Baselines -** Project planning involves creating baseline plan. A baseline plan is essentially a mirror image of the schedule of progress that is to be followed to complete a project. Performance of project is tracked using baseline.
- **Updating schedule -**In case a project is on track as per plans, then only estimation of progress needs to be made. In case a project has deviated from set plan, those activities and resources that are identified as having gone out of sync are to be updated individually. In certain situations, it is observed that projects may have certain activities that progress as per schedule, and there are some which deviate. In such cases, the desired approach is to combine the aforesaid update methods as discussed.
- **Earned Value Analysis -** As a means to integrate cost, scope and time in order to measure performance of a project and to predict the general direction a project is headed, organizations often adopt EVM. Earned Variance (EV) is applied to measure project performance during execution of project. Predictions are based after computing statistics available. Together with the analyses and predictions computed, project planners and managers can determine project performance through EVM.

For example, EVM can be based on an EV analysis conducted after fixing a project baseline and monitoring it in stages and at regular intervals. EV results are used to influence progress updates. Actual costs are also taken into consideration in EV across a time span. Schedule Variance, Cost Performance Index and Schedule Performance Index are computed in the manner discussed in following sections.

Figure 9. Window Showing Project Risks

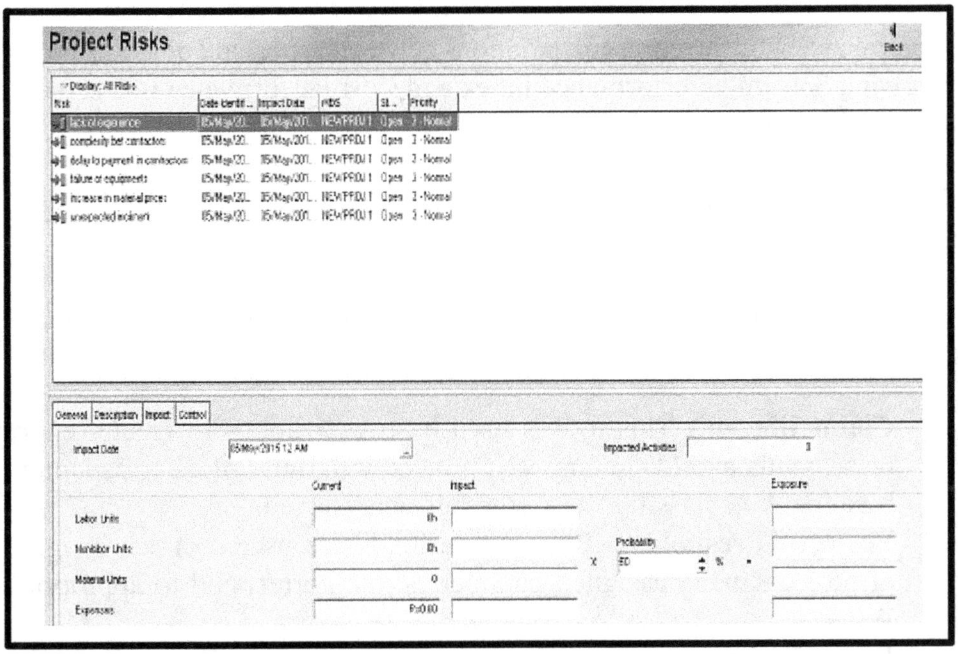

Schedule Variance (SV)

Schedule Variance represents the quantum of works completed vis-à-vis that of budgeted schedule. Schedule Variance (SV) is, thus, defined as Earned Value (EV) – Planned Value (PV).

Cost Variance (CV)

Cost Variance (CV) is the difference in planned and actual costs incurred for an individual or a set of tasks. This variance is usually represented as

CV = Earned Value (EV) – Actual Costs (AC).

Performance Indices

Performance indices represent variations between actual and planned indices involving Schedule Performance Index and Cost Performance Index. These are indicated as follows:

Schedule Performance Index (SPI) = Earned value (EV) / Planned Value (PV)

Cost Performance Index (CPI) = Earned Value (EV) / Actual Costs (AC)

- **Claim Digger -** This tool is used by organizations to create reports on select data fields in a revised project with those of a corresponding baseline. It is classified as a schedule analysis tool.
- **Project Thresholds -** Project thresholds are used to track progress of projects using assigned parameters that correspond to appropriate WBS
- **Risks** - Risks are considered to be any uncertain condition or event that has a negative effect on projects. Risks can take the form of escalation notices, imperatives, threats, warnings and any other effect that can slowdown project. Risks can also take on positive connotation. Such positive risks as and when they occur are termed as "opportunities". Documentation and analyses of risks are important as these can offer important and vital lessons for helping save costs and time in future project works. This is demonstrated in Figure 9.

Risks of an application are identified for manageability in a risk register that usually appears on a risk detail page in an application. Risks can be added and assigned to activities on activity page of PMS. As part of risk analysis, risk scoring process matrix is developed showing costs, schedules, user-defined parameters, tolerance and probabilities. Risks are grouped accordingly under these heads. Alphanumeric and numeric probability and impact diagrams (PID) are also created.

- **Issues -** Issues are problems that affect progress of project. Issues require immediate and urgent attention in order to ensure that project progress can adhere closely to budgeted costs and schedule. Issues, when identified, are tagged to a workgroup or to related project. Issues when left unaddressed or unresolved for any length of time without initiating planned action have the potential to develop into risks.

Figure 10. Window Showing Assorted Report Types

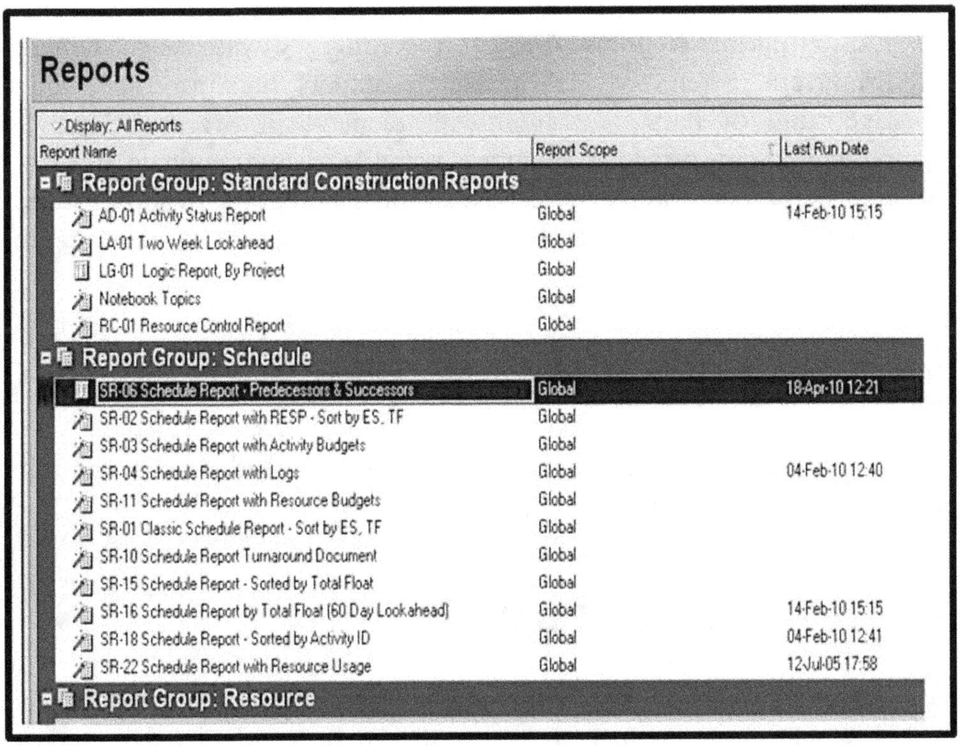

- **Issue Codes:** Issues are also codified to represent the level of intensity and degree of attention to accord to enable a quick and objective approach to resolving issues. Issue codes can have values such as high, medium, and low to reflect the nature of severity. The advantage of codifying or assigning codes to issues is that it enables searching and easy referencing to issues based on appropriate criteria.
- **Issue Forms:** Issue forms are used to gather and coalesce the necessary attributes that are required to be tracked for a given project. An issue form helps in consistent approach to identifying issues as and when these surface. Users working on a common

project need to choose an appropriate issue form to act as a template for creating any new issue detail.

- **Tracking -** Tracking windows are useful to view and monitor progress of projects using an array of layouts. These layouts can range from general project costs, labour costs, resource forecasts, allocation of resources, and such like.
- **Exporting of Reports -**Project planning software, e.g., Oracle's Primavera, often provide options to export data for distribution. Exporting of data is a significant element in overall scheme of project planning as team members are able to communicate and share information critical for project execution. The Figure 10 shows several reports that can be generated in the course of project planning and execution.

Preliminary Steps to Be Done in Updating

Project baselines must be created and saved. Once created, a project baseline is to be assigned as the primary baseline.

UPDATES TO BE MADE DAILY

1. **Start and End dates -**Activities must be marked as started when initiated. After an activity is finished, finish date is to be supplied and the activity marked accordingly.
2. **Resourcem -** After activities are created, appropriate resources are required to be added to each activity.
3. **Code -** Resources are to be assigned codes. This is usually done in an appropriate application window, or in a designated code tab.
4. **Notebook -** Often details of activities need to be recorded analogous to a notebook. From a designated notebook window offered by appropriate PMS, the relevant topic is to be selected and details jotted under the topic corresponding to desired activity.
5. **Steps -** Steps are indicative of the progress made in respect of each activity. Once all steps have been taken and are considered as having been successfully completed, the relevant activity itself is checked as being over.

6. **Feedback** -Details of resources allocated are recorded under a section often referred to as the feedback section.
7. **Work Products and Documents** - Drawings, documents detailing specifications, and any other relevant documents pertaining to corresponding activities are entered under a separate section corresponding to work products and documents.

UPDATES THAT ARE NEEDED TO BE MADE WEEKLY

1. **Threshold** - Thresholds are necessary to track mainly three types of variances. These are the start date variance, the finished date variance and the cost variance.
2. **Issues** - Values that go beyond threshold values are known as issues. In certain cases, contractors to a particular project may also add issues whenever deemed necessary.

REFERENCES

Bennett & Laurence. (1977). *Critical Path Precedence Network based management system*. Vannasta and Reinhold Company. https://www.boozallen.com/about/history/history_5 https://www.pmi.org/info/PP_OPM3ExecGuide.pdf

Kumar, S. (1990). *Computer Aided Project Management Using CPM* (M.E. dissertation). Department of Civil Engineering, University of Roorkee.

Kumar, S. (1998). *Computer Aided Cost Control of Projects* (M.E. dissertation). Department of Civil Engineering, University of Roorkee.

Lewis, J. (2002). *Fundamentals of Project Management* (2nd ed.). American Management Association.

Section 2

Managing Contracts and Materials Management With the Help of Information Technology

Chapter 4
Understanding How Information Technology Can Help in Contracts Management

ABSTRACT

Contracts management is a multi-stage process involving all parties related to a contract. It ensures that all stated objectives are met, and timelines adhered to corresponding to project execution and delivery. The fundamental premise of contracts management is to ensure quality, compliance, and cost optimization. There are several forms of contract each with its own unique features and applications. In this chapter, the authors focus on how information technology can be used to aid in supervision of contracts, materials management, survey, and design management. They also discuss the role of smart contracts and the underlying blockchain technology that is used as the basis for making smart contracts work.

Constructions contracts are extremely important not only from a financial point of view, but also consider the entire gamut of activities that would need to be undertaken to achieve completion of a construction project. It is a legally enforceable and binding agreement between a construction company and its client, and contains therein related details and costs of the project itself. Construction contracts are can involve from simple renovation jobs and upwards to expensive and complex contracts. The types of clients that use construction contracts are commercial and residential. Depending on

DOI: 10.4018/978-1-7998-5291-9.ch004

client requirements, contracts are prepared accordingly. In general, residential type of construction contracts involves project scope, schedule of work and payment details. A project scope defines what kind of construction work is included in contract. Both parties to a contract must agree to abide by the kind of work that needs to be done. Details of project costs and dates of payments are scheduled and recorded in payment detail section. Payments are paid against percentage of total work completed of a project. It is generally an accepted practice that a deposit not exceeding five percent of the total costs is disbursed by client to the construction company before project work commences. Subsequent payments are disbursed corresponding to scheduled completion of sections of project work defined and agreed upon by the parties concerned. As in every other sphere that Information Technology has today permeated, construction contracts management, too, benefits from inclusion and influence that Information Technology has in terms of scheduling, managing and tracking construction activities. Efficient management of construction contracts management is important for maintaining quality and consistency at all times throughout life-cycle of project.

WHAT IS CONTRACT MANAGEMENT?

The significance of contract management is immeasurable. It is a multi-stage process that progresses along the lines of contract schedule. Its foremost purpose is to act enable all parties involved in a contract to act and deliver stated objectives as specified in terms of contract. The chief objective of contract management, however, is to ensure that all stated objectives as detailed in contract are adhered to at all times and, thus, ensure efficiency and resource optimization (Gustavo, Kasun, 2013). The underlying principle behind contract management is to seek quality and cost optimization throughout project execution.

The aims and objectives of contract management can be detailed as follows:

Effectiveness

The primary objective of contractual management is to execute a contract successfully. In general terms, this involves the following:

1. Timely delivery of goods and, subsequent, installation and/or consumption in the stated manner as per contract details.

2. Proper execution of services as per scheduled timeline.
3. Scheduled completion of all related civil works.

Efficiency

In order to carry out all activities efficiently, the following aspects need to be carefully considered and observed throughout the entire life-cycle of project:

1. Budget (costs)
2. Duration
3. Quality monitored across multiple parameters

At all times, it is to be borne in mind that striving for efficiency without due consideration to the human factor may result in plotting unrealistic goals, or even compromise quality and act to the detriment of the project itself.

DIFFERENT TYPES OF CONSTRUCTION CONTRACTS

A few of the more prominent types of contracts used in the construction business are the following:

1. Lump Sum / Fixed Price Contracts
2. Cost Plus Contracts
3. Time and Material Contracts
4. Unit Price Contracts

The choice of contract depends on the primary stakeholder's risk appetite.

Lump Sum or Fixed Price Contract

Lump Sum or Fixed Price contract – as the name suggests, is a fixed price agreement against which a contractor agrees to complete a task assigned as per terms of the contract agreement entered by and between the contractor and the remaining parties to the contract. Such contracts can only be changed upon mutual agreement and concurrence of all parties concerned and subject to either additional work falling outside the scope of contract and against additional payments, or excising a specified amount of work from the original agreed upon scope of work simultaneous to reduction in the price (Chen,

Okudan, Riley, 2010) proportionate to the work that the contractor is no longer obliged to execute. Such types of contracts are usually appropriate when there is a clear scope and pre-defined schedule that has been mutually agreed and reviewed in advance. The benefit of Lump Sum or Fixed Price contracts lies in the fact that the primary stakeholder is aware of the costs involved and can predict costs to be apportioned at every stage of the construction work (Chen, Kamara, 2011). So long as there are no significant changes to the contract nor any disputes arising out of the project work being undertaken, the total cost of such contracts are usually capped by the contract prices by which they are defined in the respective contracts. While there are not many drawbacks with such contracts, it remains, nonetheless, extremely important for all stakeholders and parties to the contract to clearly define the scope of work thoroughly prior to commencement of work (Chow, Thomas, 2007).

Other types of contracts such as Cost Plus Fee or Time and Materials Contracts can either be cheaper than Lump Sum or Fixed Price contracts, or more expensive depending on the actual volume of work involved and in case of any liabilities or disputes arising. For such reasons, Lump Sum or Fixed Price contracts are considered more convenient to be used for construction projects.

Cost Plus Contracts

In Cost Plus Contracts, the primary stakeholder or owner of the project agrees to pay the whole cost of materials and labour required to execute the project in addition to paying the fee to the contractor involved for its overheads and profits. Such contracts are widely used where the scope of the work can be ambiguous or nebulous with no fixed timeline or scales of labour, material and equipment known in advance. In such contracts, complete and detailed records of all materials and resources deployed by the contractor, along with durations involved, need to be meticulously maintained. Contractor's fees that have been negotiated in advance as a percentage of resource costs need to be considered to cover its overheads and profit margins. Costs involved are usually classified under direct or indirect costs.

There are several types of Cost Plus Contract. The more common ones are listed below:

1. Cost Plus Fixed Percentage Contract: A percentage of costs account for compensation to be paid to the contractor.

2. Cost Plus Fixed Fee Contract: The contractor is paid a fixed sum irrespective of project costs. The primary stakeholder or project owner is obliged to reimburse actual costs incurred by contractor regardless of the amounts involved, and in addition to a pre-determined and negotiated fee independent of the costs incurred.
3. Cost Plus Fixed Fee with Guaranteed Maximum Price Contract: A fixed sum of money is agreed upon in advance as compensation with the total project cost not to exceed an upper-bound limit.
4. Cost Plus Fixed Fee with Bonus Contract: While compensation is based on a fixed sum, a bonus is agreed upon in advance, to be paid in case the contractor can finish the project either ahead of schedule or within budget, or both
5. Cost Plus Fixed Fee with Guaranteed Maximum Price with Bonus Contract: Compensation is based on fixed sum with the total project cost not to exceed a mutually agreed upon upper-bound limit. A bonus is to be paid in case the contractor finishes the project either ahead of schedule or within budget, or both.
6. Cost Plus Fixed Fee with Arrangement for Sharing Any Cost Savings Contract: In such types of contracts, compensations are fixed with any cost savings shared between the owner and the contractor.

As the name suggests, Cost Plus Fixed Fee construction contract is rather more predictable than Cost Plus Fixed Fee Percentage construction contract because in the case of the former, the contractor's fees for overheads and profits are mutually agreed upon in advance. However, the Cost Plus Fixed Percentage Construction Contract offers contractors with an incentive of not keep costs of production low as fees would doubtlessly increase with associated costs of construction incurred. It is for this reason that Cost Plus with Guaranteed Maximum Price Contract is arguably attractive as it limits the owner's financial exposure should there be a runaway increase in the costs of construction of project.

To help motivate contractors to complete projects within or before agreed schedule, quality and cost constraints, Cost Plus Construction Contracts are frequently embellished with bonus contracts, contingency allowances or even grouped with cost savings contracts.

Unit Price Contracts

Unit Price Contracts are straightforward in that the final project price is contingent upon bills of materials that are required to be prepared to carry out the work. The key feature of such contracts is that these involve highly accurate picture on the different types of items involved, and not necessarily the respective quantities. These contracts are found suitable in case of excavation projects.

Time and Material Contracts

Time and Material contracts are chosen in cases where the scope of the project cannot be made distinct in advance. In such cases, parties involved need to agree on hourly / daily rates of work and including any additional expenses that may incur during construction activities at a later date. The project owner may adopt an approach to seek an upper limit to his exposure to cost escalations should the contractor exceed time and budget limits.

CONSTRUCTION PROCESS

At the beginning of a contract, a contractor may often be required to deliver documentary evidence in support of insurance, performance bonds, provisions for preliminary contractual programmes necessary for the owner to accord access to the physical location or site to the contractor.

Allowable Rates

Even before actual work can begin, a contractor is expected to make thorough preparations to take care of resources and the methods of work that are required for successful execution of the project. During this stage, certain lump sum amounts are needed to be paid to kick-off productivity requirements. The contractor usually relies on a history of similar productivity rates to determine the actual quantum involved. "Allowable rates" or "lump sum amounts" are defined as the time and costs involved that a contractor has expected to account for labour, equipment, transport, profit, etc. during execution of the project. The contractor is also expected to assume the manner in which the project work is to be accomplished.

Resourcing the Project

Planning is an essential ingredient to a project. To effectively construct a plan, a contractor needs to identify from the drawings, items and quantities that would be required during construction works, including resources and equipment. Thus, the contractor is also usually able to determine the profits that it can make from the project after its successful completion. Resource planning is usually the culmination of works programme and the rates at which the contractor can execute the project. Needless to mention, of course, work programmes are extremely essential for project execution.

WHAT IS CONTRACT ADMINISTRATION?

Administration is defined as managing the business of a construction contractor. The management of business affairs at a location of the contractor is said to be site administration. While site administration and site management are extremely important and akin to "nerve centers" of operations, most of the time these two facets are often ignored leading to loss in production and poor quality of work. An efficient and effective contract administration can go a long way in minimizing claims and disputes arising out of variations, final accounts, design specifications, etc.; and, thus, lead to reduction in costs of production and construction.

Record Keeping

General

Construction projects being complex businesses to own, operate and manage, record-keeping assumes a vital aspect to ensure that events of what has taken place at every stage can be recalled. Records may be kept in several forms ranging from site diaries, drawing registers, notices, correspondences, site instructions, site photographs, contractual documents, health safety and environment documents, etc. Proper filing is necessary so that such records can be retrieved at a later date. Record keeping and filing are both necessary for a contractor to quickly look up any important event for its reference during tendering or commencing any new project. Inarguably, production costs are lowered, and improvements can be made in terms of production and quality.

Site Diary

A site diary is an important document that records logs of activities accomplished at a site each day. It is considered as a chronological log of events that record the kinds of problems faced, instructions received from primary project owners, drawings received, and production milestones that are reached. Since, every project has its unique signature pattern, a site diary correspondingly changes for a project, accordingly. A site diary needs to include, in certain cases for contractual reasons, information on day, date, rainfall recorded, schedules of personnel, schedule of equipment, site instructions received, drawings received, issues causing delays, rectifications of work done, site meetings held and the next scheduled meeting, etc.

Drawing Register

A drawing register is a must for all project sites as it is used for recording changes made to drawings, and the dates when such changes are made. Each drawing is identified with the help of the drawing number, revision number, and the associated date. The purpose of a drawing register is to record when and what changes were made that may have influenced the way the project works have progressed.

Written Notices, Correspondence and Site Instructions

Correspondence is a vital to convey and confirm discussions held between stakeholder of the project at site. Records of what has been discussed, and in under which context, is crucially important. While written notices are somewhat similar to correspondence, the former differ in that references are usually drawn to particular clauses in contract.

Site Photographs

Keeping a record of site photographs and photographing works conducted at site are necessary for maintaining a chronological sequence of events. Site photography becomes essential when ascertaining record of conditions that may have existed at site on any given date and time. With the advent of digital photography, it has become convenient and relevant to store date and time, location, reason for photography, name of photographer, names of relevant people who may appear in photographs, names of subcontractors in photographs, etc.

Contractual Documents

A vital document for any construction project is the contract itself. This details rights, responsibilities and obligations of the parties and stakeholder of the contract. Copy of this document should be available to all site personnel for reference and kept in a secure and safe location. Documents that can have contractual reference may include, but are not limited to, main contract, subcontracts, labour contracts, plant & machinery hiring agreements, delivery notes, programme schedule, minutes of meeting held at site, notifications for inspection, completion certificates, health and safety documents, daily labour sheets and materials, reimbur sables, etc.

Understanding the Contract Documents

A basic level of understanding is a primary requirement for project execution. Documents are defined and described in writing to aid in the construction process. Graphic documents are also prepared by designers to facilitate in communication and carrying out administrative oversight of construction-related activities.

Various Documents Constitute the Contract Documents That Are the Basis of the Contract

Geotechnical data and surveys form part of other documents that are required for reference. There are other documents that may no longer be applicable once agreement is signed and contract finalized. These documents can range from procurement solicitations, procurement instructions, related forms, bid security, etc. However, an understanding of how these documents are prepared can help in extracting potential benefits from such. Following abbreviated descriptions of documents used in construction projects are listed:

1. Articles of agreements, or deeds that are also known as General Conditions of Contracts.
2. Contract drawings.
3. Bills of Quantities.

4. Project Specifications.
5. Specifications.
6. Schedules of work.

Construction Contract Documents

Contract documents are usually listed and serialized in a construction related project agreement. These form a legal trove of documents vital for reference and for describing the nature and scope of the contract. Contract documents describe the proposed nature of work, allocation of resources, manpower, and design drawings. As explained earlier, contract documents may include both graphic and written matter. These can encompass drawings of various types ranging from contract, coordination, record to resource drawings, electronic models, specifications, etc.

RELATION BETWEEN CONSTRUCTION PROJECT MANAGEMENT AND CONTRACT MANAGEMENT

One of the vital topics understands the relationship between construction project management and contract management. Contracts form the lynchpin in construction project management, thus, rendering such as one of the most vital set of documents that govern the scope, nature and ultimate outcome of a construction project. The information and data that are involved in contracts and ultimately construction project management, if effectively harnessed can provide an edge over business competitors. It, therefore, becomes an automatic necessity to develop an information management system through which gathering, processing, storage and transmission of data can take place steadily and swiftly.

THE NECESSITY FOR INFORMATION TECHNOLOGY IN CONSTRUCTION CONTRACTS MANAGEMENT

Construction engineering and construction contracts being unique, contracts management itself must have the following features:

- Since construction projects are a gradual process spread over long durations, frequent changes and adjustments need to be considered.
- Contract management has to cater for receiving continuous bids until project completion.
- As most contract projects are of considerably high value, the economic influences on engineering tasks and activities need to be monitored closely.
- Contracts being extremely complex activities, ensuring accuracy are rigorous activity. Assimilating, analysing and implementing information and parameters contained in assorted contract documents are extremely essential. The vast multitude of files and data produced can only be effectively managed through an appropriate information management system supported by modern Information Technology tools.

WHAT IS CONTRACT MANAGEMENT SOFTWARE?

During the days when Information Technology had yet to make deep inroads into modern day life, it was common for contracts to be kept in physical file cabinets or maintained as simple spreadsheets that contained different agreement dates and related entries noted. While this was simple and involved no particular amount of additional investments or maintenance costs, it gradually led to a whole lot of problems. Physical contracts could be filed randomly, physically abstracted from filing cabinets and not returned at all. Further, a change in contract manager or an audit exercise can overwhelm the concerned officials in charge with so many records as to lead to an almost total loss in efficiency and untraceable records. In turn, the resulting delays and absence of proper documentation can cause unexpected cost escalations, trigger unintended renewals or may even cause certain intellectual property (IP) rights to go invalid.

To address these issues and mitigate the human risks posed, organizations have resorted to implementing contract management systems. It is often claimed that contract management systems exist in the market by their hundreds all offering services that aim specifically at increasing auditability, reduction in long-run costs, savings in time, etc.

THE APPLICATION FOR INFORMATION MANAGEMENT IN MANAGEMENT OF CONSTRUCTION CONTRACTS

It cannot be overstated that without a working information management system in place, the management of construction contracts can falter and, thus, lead to serious quality related issues. There are a number of reasons for unsatisfactory information management scenarios. The primary reasons involve low quality of training among contracts management personnel and external environment.

The Impact of the External Environment

External factors such as change in engineering designs, interference from the executive, payment arrears, etc. are matters that serious interfere and impede in contracts management in construction engineering processes. The ill effects of such can be remedied through information management that can help respond quicker to the affected engineering changes in the construction project.

The Restriction of the Contracts Management Personnel Quality

In many instances, contracts management staff who have been exposed to traditional methods of approach to contracts management may balk at adapting to changes required to adopt information technology in contracts management. In such cases, the concerned staff should be encouraged to adopt these modern IT tools to seamlessly integrate information flow and data processing.

THE FEASIBILITY FOR INFORMATION MANAGEMENT IN MANAGEMENT OF CONSTRUCTION CONTRACTS

The objectives of Information Systems are multi-fold:

- Contracts supervision, materials management, survey and design management becomes more systematic.

- Manually sifting through huge volumes of contract data can be automated through Information Systems. Thus, contract managers can be free to look closely at other significant areas of interest where due attention in advance can help reduce cost-overruns.
- Contracts can both be prepared and searched for as per FIDIC Conditions of Contract guidelines. Composition of contracts can be rapid and be more relevant to the project concerned.
- Information system function

Figure 1 shows the organic nature of system functions in contract systems management.

- The Contract Document Management module collects and organizes materials and all other relevant information for search and documentation. Data in this module can rage from project bid notices, tender documents, BoQs, technical data, etc.
- The Terms of the Contract search module assists in searching for terms of contract based on either conditional or unconditional search parameters.
- The Contract Management Module involves data on costs, schedules, and other relevant information on quality and other operational aspects.
- The Contract Change Management submodules exist to assist in generation of lists of changes and applicable costs to factor in while computing the total amount of investment changes made. Claims management and processing of related queries are part of this module.

SOFTWARE OVERVIEW

While most of are well-versant or at least familiar with Microsoft Excel and similar other popular open source spread sheet alternatives that support data presentation and calculation in rows and columns, there is rather a steep learning curve involved in understanding databases that are considered a 3D version of spread sheets. Some have even compared databases to Rubik's Cube. Construction professionals have often found much use for databases such as MS Access and MS SQL, Oracle's MySQL and its paid counterpart. Once mastered or at least familiarized, databases yield far more productive gains than spread sheets. Construction contract administration software programs rely on databases to manage information and display it in different forms such as charts, tables and templates. Displaying data from databases are usually less complicated than spread sheets when considering huge complex data sets. Such software are usually categorized under databases, notification databases, contacts and forms, and cloud. Each software has its own special trait with the last type referring to the latest in cloud computing services and the Internet.

Figure 1. The system functions

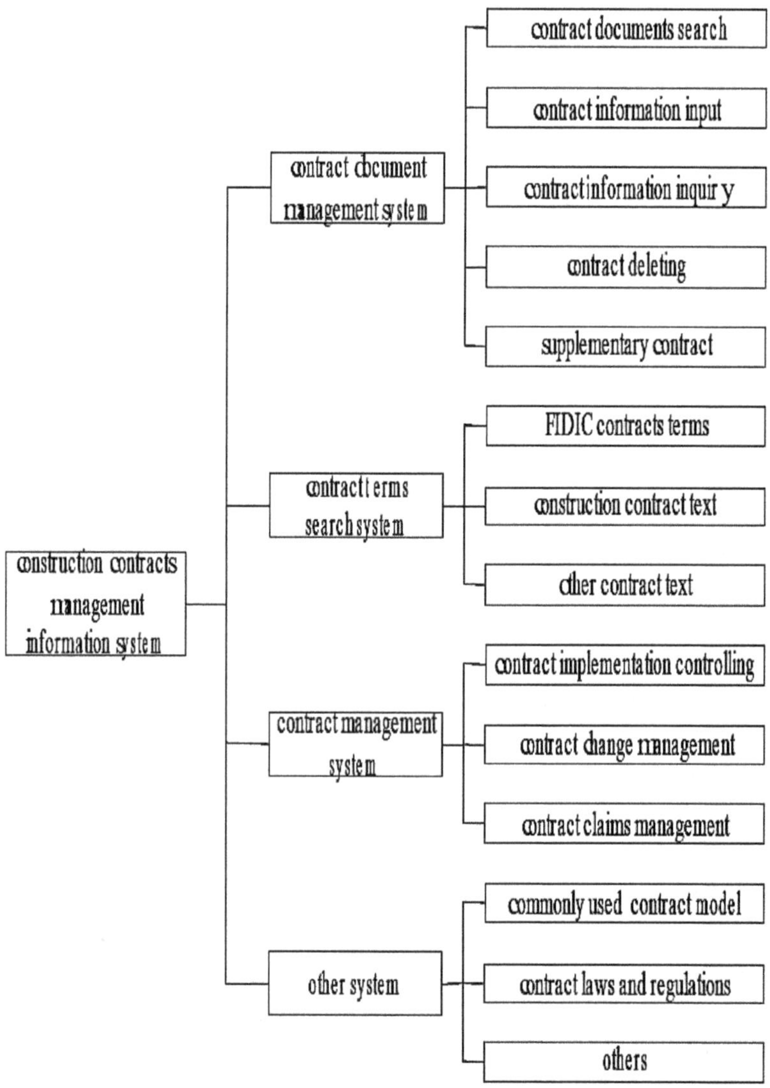

SMART CONTRACTS IN CONSTRUCTION INDUSTRY

Despite the profusion of software to manage contracts and other contract-related paraphernalia, the concept of smart contracts began to take root even as early as in 1994. First expounded by Nick Szabo, the concept of embedding smart contracts in physical properties to transform such into smart properties was quick to gain focus. Construction industry suffers significant losses in terms of business failure to meet committed payment schedules or even outright refusals to pay even though the contracted job may have long been completed satisfactorily and in terms of contract agreed by and between the principal stakeholder and contractors. It was postulated that blockchain technology and cryptocurrencies can help smart contract use to mitigate most of the problems that ail the construction sector. The fundamental idea behind blockchain and smart contracts is that funding is released only and immediately upon fulfilment of mutually agreed contractual terms and conditions. Software coding ensures that trust based on immutability of data, helps in facilitating automatic payments when stipulated conditions are either met or exceeded. Software code and smart contracts can, thus, reduce or even negate the influence of third party legal teams who are often called on to settle disputes within legal framework. Payments can be made either using Bitcoins or other forms of digital currency. Compared to traditional contractual ledgers, smart contracts have several advantages: Smart contracts automatically release funds once an agreed upon task or activity is completed. This reduces any lag involved over traditional contractual ledgers. Since, smart contracts run on computer software and logical precise conditions built in, execution of smart contracts is automatic and requires no human supervision or intervention. This, further, helps in reducing time required in operational aspects. The decentralized nature and immutability of blockchains on which smart contracts are based, make it almost impossible to tamper and cause debilitating effects to trust-based relationships among all parties involved.

APPLICATIONS OF SMART CONTRACTS IN CONSTRUCTION INDUSTRY

Smart contracts can have a wide range of applications in construction industry, beginning with procurement phase. Where multi-contracts are involved, procurement process becomes a major time-consuming exercise. In cases where buyers and sellers are negotiating for business for the first time, the factor of trust becomes crucially important. It is here that smart contracts can replace middlemen, banks or even notaries by virtue of immutability and decentralized nature of database consisting of blockchain. Smart contracts can even take the place of letters of credit. A letter of credit (LoC) is a document that establishes trust through banks. It is an ironclad guarantee that a seller shall be paid after all contractual obligations are met .

The following Figure 2, demonstrates how a party importing machinery can establish trust with another party. As would be evident from the figure, the party importing the machinery opens a LoC with bank following agreement of contract and terms and conditions. While it is not the intention of this book to explain LoC in detail, it is nonetheless important to develop on the point that opening LoCs is expensive and also liable to be time-consuming in case of disputes should these arise.

Figure 2. Letter of Credit Document Process

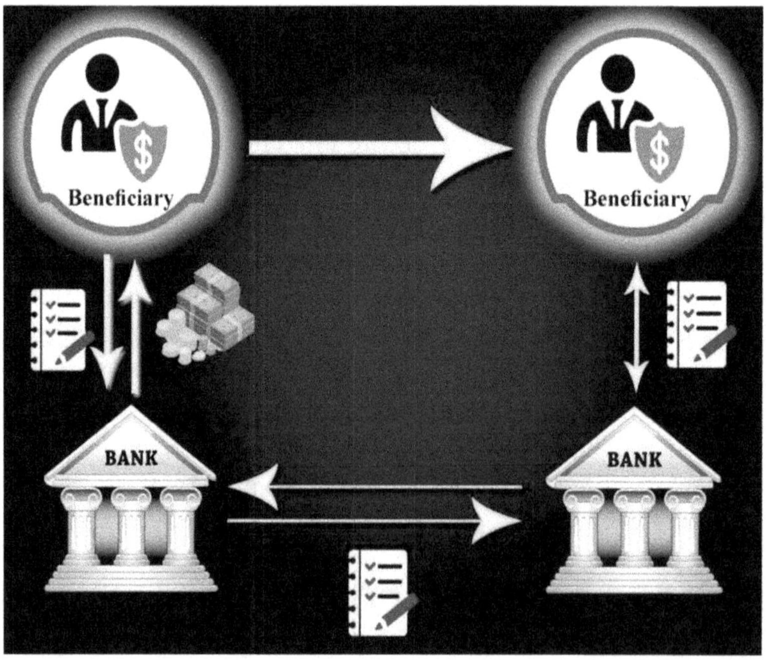

Smart contracts have the potential to replace expensive LoCs easily. Transparency, promptness, trust and immutability combined make smart contracts a far more viable and reliable alternative than traditional means of establishing trust via banks.

HOW SMART CONTRACTS CAN BE USED IN THE CONSTRUCTION INDUSTRY

A contract is usually signed on the basis of the project owner putting up a set of specifications backed up by relevant drawings and other associated documents, with a tender called to invite interested contractors to bid for the job. Based on a simple letter of intent at times, or often not even that document at all, a contract is entered by and between the parties concerned to complete a project in exchange of a certain amount of money.

Usually a contractor will then look at the scope of work closely and then hand out subcontracts to other subcontractors to perform related activities within their individual expertise. Irrespective of such arrangements involved, once a particular section of work is done, payment will need to be paid in terms of the contract and subcontracts specified. It is in such scenarios that smart contracts can prove to be useful as the principal stakeholder, or owner of the project, can directly embed digital currencies into the contract that are to be released only upon fulfilment of a set number of conditions. Since, payments are automatically integrated and included in the contract before a project commences, contracts and subcontractors can be assured of payments made once individual works within their respective scope of works are delivered.

Smart contracts can also be interlinked to form a web of payments. As a whole, smart contracts are beneficial for contractors and suppliers who often suffer losses as a result of payments withheld or from insolvency declared by any party or parties higher up in the chain of work.

HOW DO SMART CONTRACTS WORK?

Smart contracts are dependent on a technology like the one on which Bitcoin operates. A distributed system of ledger for accounting transactions, known as blockchain, is the basis on which smart contracts are formed. In a blockchain records in each transaction are stored in blocks. These blocks have the following characteristics. It may be borne in mind that the concept of blockchain itself is that of a distributed database where any change affecting a particular block is propagated to all parties that have copies of the same database.

Figure 3. A Simple Model for Smart Contracts

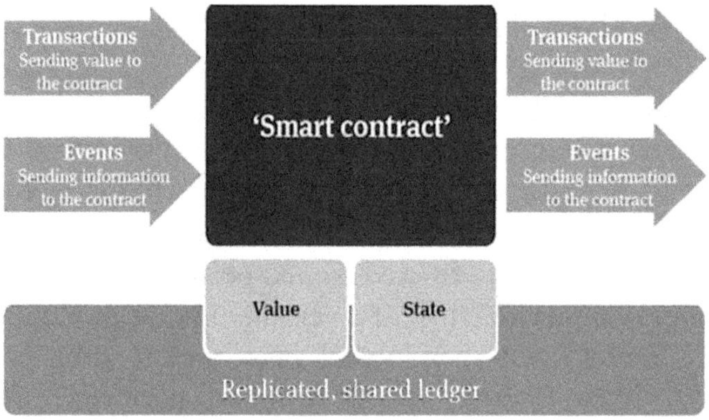

- **Hashing:** Every block has a hash that is unique. Every block references to the previous block in the chain.
- **Transparency:** If data in any block in the chain is altered, this becomes immediately apparent to all members of the blockchain. Changes can be tracked as the block's hash will alter and, hence, subsequent blocks will no longer relate to the changed block.

Blockchains can either be public, i.e., open for all to inspect but controlled by no one in particular, or these can be private as in a virtual private network. Since, blockchains operate on a distributed basis, records and ledgers of all transactions are replicated in full in each participating node or of each participant's computer. As each participant has a personal copy of the entire blockchain ledger, every transaction in the blockchain can be traced. Smart contracts operate on blockchain principle and function on distributed basis. It is important to note that like blockchains smart contracts rely on cryptography for authentication and verification.

A typical functioning of smart contract can be represented as shown in Figure 3.

BLOCKCHAIN-ENABLED CONTRACT MANAGEMENT

One of the more significant advantages of smart contracts that rely on blockchains are that these do not require external influences to act and execute on their own. Parties to the contract are aware that on fulfilment of all parameters, the contract itself shall execute on its own. Smart contracts typically are reliant on special kind of software to automate tasks that would otherwise occupy a whole lot of time in processing relevant and all associated paperwork. Not only does this render services of middlemen unnecessary, it directly helps in accelerating transactions, cutting down costs and improve trust among parties to the contract while maintaining transparency and traceability at all times.

Blockchain allows smart contracts to be enabled between parties, and executed on an Information Technology controlled environment regardless of the scale and complexity of the contract involved. The Figure 4 sketches the process of execution of a smart contract, and describes the manner in which transfer of funds takes place from one party to another through blockchain.

1. In this step, a transaction is requested. The request is challenged through to a peer-to-peer (P2P) network. These networks that can be either private or public then verifies the identities of the parties involved. A transaction may involve contracts records, building models, drawings, etc.
2. In this step, the data of the transaction is represented in a Block.
3. The Block then gets transmitted and distributed across the P2P network.
4. The P2P network conducts validation of transaction via a Consensus-based code.
5. The block can then be added to blockchain that is designed to be an unalterable log of transactions.
6. Since each node in the network has the exact same version of the chain, any attempt to commit fraud is easily deterred and can be nullified. Changes that are not authenticated are not allowed to propagate.
7. Soon after a transaction is completed, the associated contract is fulfilled automatically.

Smart contracts allow trust to be propagated across P2P network community rather than being centralized at a fixed location.

Figure 4. Process of execution of a smart contract

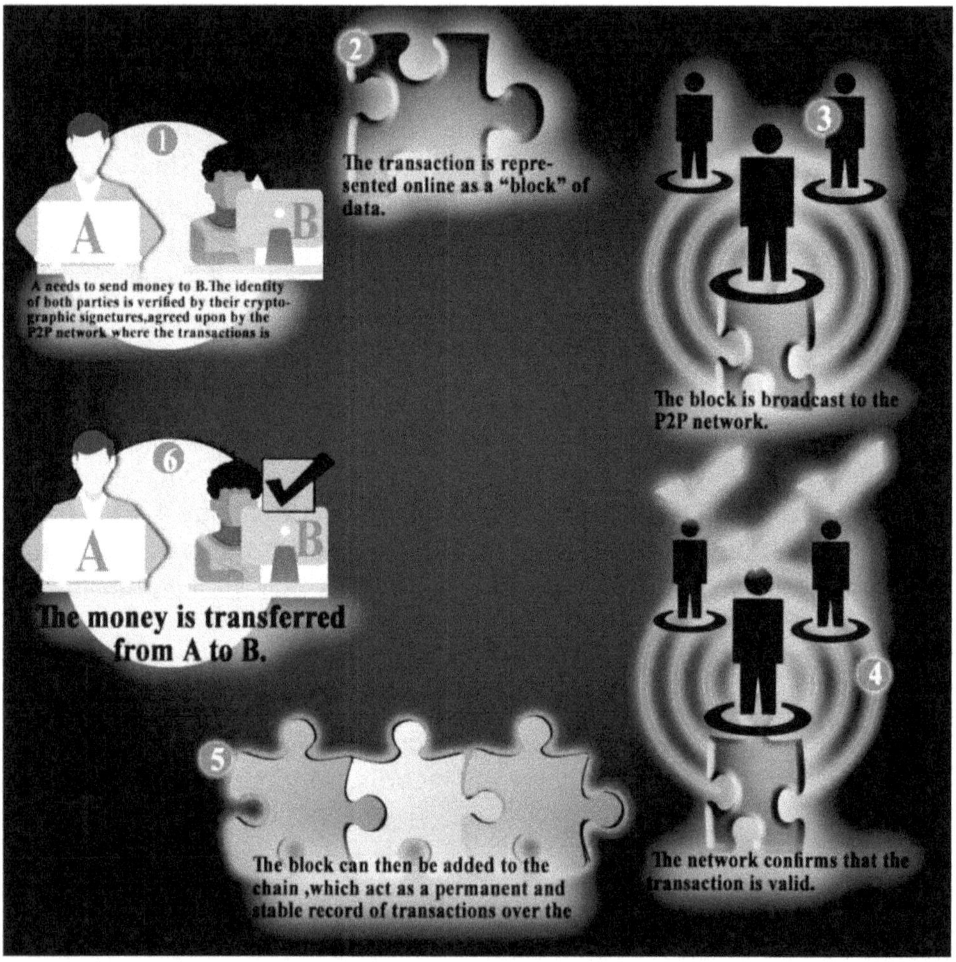

It is important to bear in mind that for a blockchain monetary system to function, a transaction must be able to confirm that:

1. The transaction initiator has the funds to back up the transaction.
2. The funds have been sourced through valid means.
3. The result of the transaction shall be that the recipient shall be identified by everyone as the possessor of the funds being transferred through the transaction.

4. In case of successful transaction, the transaction initiator shall no longer be able to present as being in possession of the funds that have been transacted.

A contract is programmed with the above factors. A P2P network describes a series of if/then dependencies which mandate that if all the four conditions are met, then the funds transfer is allowed to go through.

Depending on the kind of access required, blockchains support several configurations. A public blockchain grants access depending in a Proof of Work (PoW) to be performed. On the other hand, a private blockchain grants authority over transactions to a single owner instead of decentralizing ownership depending on PoW. A hybrid or consortium blockchain is a combination of private and public blockchains that may involve several collaborating organizations and entities.

Implementing blockchain in construction sector can aid in several ways as described in the following sections:

Payment related and cash-flow issues can be mitigated by integrating smart contracts. It is often a common and unfortunate practice of withholding payments that results in costly litigation and disputes. A smart contract can alleviate such problems by embedding funds into the contract itself. Thus, small subcontractors and suppliers can be indemnified against insolvency and/or other cash fund crises. Milestones can also be built into smart contracts wherein funds are automatically released on completion of phases of the project.

Smart contracts leave less room for ambiguity. Thus, these are predicable and can be used for contract registration, monitoring, updating through automatic processes that are also tamper-proof in nature.

Blockchain and smart contracts hold forth the possibility of reducing reliance on legal teams for reliance for enforcing contracts. Blockchains operate on the premise of immutable trust and non-editable format. As a result, parties involved in smart contracts can be more productive and help reduce costs.

REFERENCES

Chen, Y., & Kamara, J. M. (2011). A framework for using mobile computing for information management on construction sites. *Automation in Construction*, *20*(7), 776–788. doi:10.1016/j.autcon.2011.01.002

Chen, Y., Okudan, G. E., & Riley, D. R. (2010). Decision support for construction method selection in concrete buildings: Prefabrication adoption and optimization. *Automation in Construction*, *19*(6), 665–675. doi:10.1016/j.autcon.2010.02.011

Chow, L. K., & Thomas, S. (2007). Expectation of performance levels pertinent to consultant performance evaluation. *International Journal of Project Management*, *25*(1), 90–103. doi:10.1016/j.ijproman.2006.08.007

Gustavo, E. A., & Kasun, N. H. (2013). IT based system for construction safety management and monitoring: C-RTICS. *Automation in Construction*, *35*, 217–228. doi:10.1016/j.autcon.2013.05.007

Chapter 5
Information Technology in Materials Management

ABSTRACT

Ensuring proper supply and timely delivery of materials during every stage of construction is crucial for overall project success. The task of managing materials is difficult and requires considerable effort on the part of engineers, project planners, supervisors, logistics teams, and storekeepers. This chapter explains the problems associated with materials management and how information technology can help in solving those issues. The authors discuss the use of RFID and the role that this technology can play in tracking shuttering components, such as Mivan, and many similar high-value raw materials and even smaller items. They also discuss how bar and QR codes can help in materials management.

INTRODUCTION

Construction projects are dependent on assorted and large volumes of raw materials. Controlling, regulating, procurement, storing, deploying, and disposing unused raw materials are essential processes that constitute bulk significance for construction companies. Management of materials, or Material Management, is critical to success of any construction project. The management of materials can throw a number of challenges to engineers, project planners, storekeepers beginning with material shortages cropping up at unexpected times, delays by suppliers, rise in rates of supplies, wastage and damage, to

DOI: 10.4018/978-1-7998-5291-9.ch005

lack of holding space for storing raw materials. Maintaining a paper-based trail to keep track of inventory and managing materials (Arnold, Chapman, 2004) prove not only difficult and cumbersome to maintain, but also prone to errors, delays, insufficient information, and in general an inefficient process unsuited in modern construction sector where time is of the essence and cutting down on wastage could mean survival for the construction company itself.

Information Technology has been harnessed to aid in construction materials management (Harris, MacCaffer, 2001). Sophisticated IT solutions such as Radio Frequency Identification (RFID) for tagging materials, bar coding, and wireless communications are already widely in use. An appropriate IT solution (New Strait Times, 2007 targeted at making materials management in construction projects effective is considered an integral part in any construction materials management planning exercise. It is already evident through research and practical implementation that a well-planned and advanced materials management solution can help improve inventorying and tracking of materials at a construction site. That materials management is of immense significance in boosting productivity in construction sector, is undeniable. Materials management takes into account that all materials should be carefully accounted for at all stages and phases of construction. Improper management of materials can and does lead to deterioration in quality of construction besides slowing down construction works and adversely affecting budget and schedules. It, therefore, becomes critical that the right amount of raw materials of appropriate quality standards are available in time at every stage and process of construction. The cost of materials, too, is also a part of consideration while engaging in materials management processes. While many organizations often still rely on manual methods for keeping track of materials (Kasim, Peniel, 2011), this leads to wastage and cost overruns. Paper-based techniques are ill-suited in modern construction environment.

Information Technology has the capacity to improve and advance materials management at construction projects. Already software applications form the core support of all business processes in construction companies big and small. Software applications are known to facilitate in collaborating, sharing knowledge, and procurement practices in many construction companies. With the aid of IT, it is possible to extend product procurement (Sarshar, Isikdag, 2004) features involving direct or indirect purchasing, making payments electronically, and managing movement of materials. Such can not only eliminate paper work, but also aid in lowering procurement and operational costs while cutting down cycle times and reducing waste. With the help of Information Technology, construction companies today, are not only able to pursue an efficient course in materials management, but also formulate transparent and often accurate picture of cost estimation for raw materials. Software such as Microsoft Excel are considered invaluable tools in the trade. Significantly, however, the use of relevant IT tools and equipment such as bar codes, QR codes, RFID, and supported by wireless communications are yet to make a deeper foray into the world of construction sector at project sites where these would be able to better serve planners and engineers to stay on time and within budget.

MATERIALS MANAGEMENT

Materials Management is a coordinating function that controls and regulates the processes of planning and flow of raw materials. In finer detail, materials management can be considered to be a well-defined clearly laid out approach involving planning, scheduling, purchasing, delivering, handling, and storing with the aim of reducing waste and meeting objectives of requirement of resources within the overall project schedule. Materials management is a critical exercise that boosts productivity. As such, the efficient use of materials and the management thereof can have a considerable impact on the overall operational productivity of a construction company, and also help prevent delays and cost overruns.

To better understand materials management, it is essential that the following processes involving planning, procurement, logistics, handling, stock and waste control be discussed.

- ***Planning*:** Determining quantities of materials required, ordering and scheduling of materials are all part of materials planning stage. This stage is important as it properly executed planning process has an effect on the overall productivity and profitability aside from aiding in adhering to project timelines. Productivity and costly delays may result if material planning process is not carefully carried out.
- ***Procurement*:** In this process, materials and services are ordered and purchased from outside organizations and agencies to support construction company's own requirements related directly to production, logistics, marketing and sales. Coordination and material scheduling are vital tasks that need to be actively monitored and carried out to ensure that materials are on hand at every scheduled stage.

- ***Logistics*:** This involves planning, scheduling and regulating the flow and storage of all goods and materials right from raw materials through to finished products. It has been observed that this process if not executed properly, results in costly delays and cost overruns.
- ***Handling*:** Material handling is an important aspect in construction sector. This has an important effect on how materials are processed and handled along with manpower utilization for processing materials at various stages of production.
- ***Stock and Waste Control*:** Stock control is a technique designed to cover and ensure all items are at hand when required. Stock control covers raw materials, processed materials, assembly components, consumable stores, general stores, maintenance materials and spares, work-in-progress goods and finished products. In the absence of stock control, delivery of a major portion of construction materials will not be effectively rendered leading to cost overruns and delays.

It has been observed that construction activities often are not efficient and produce huge amounts of waste. This problem is considered critical by many observers in the construction industry and is also recognized as one of the key issues that lead to loss in productivity. An effective material planning process set can lower waste produced thereby contributing directly to productivity and profits. There are, however, certain hurdles that beset the construction sector, and which are discussed in the following section.

MATERIALS MANAGEMENT PROBLEMS

The key problems that construction companies face in materials management are listed as follows:

- Unavailability of proper storage space at site.
- Moving and installing tower cranes at desired locations.
- Materials logistics.
- Insufficient space for loading and unloading areas.

These along with several associated issues, and covering materials management process as a whole, can be addressed through implementation of appropriate Information Technology systems and tools.

IT IMPLEMENTATION IN MATERIALS MANAGEMENT

Information Technology (IT) is generally deemed to mean the use of electronic machines and digital devices that are controlled by software programmes to achieve a wide range of tasks that essentially constitute display, processing, storage and transfer of information and data. Users involved in materials management processes have traditionally relied on Microsoft Office software, and a range of computing hardware that includes laptops, desktops, mobiles phones, tablets, portable data terminals and, the now defunct, personal digital assistants (PDAs). However, as demand for faster processing and larger information storage has grown, so too has the levels of IT sophistication turned to more intricate and higher forms. Nowadays, construction projects are turning to wireless communications, bar codes, radio frequency identification systems (RFID) for tagging materials to render easy identification and delivery of materials from one location to another. Inventory management, thus, has been redefined and brought to a whole new level where it is now possible to track movement of materials in almost near real-time. The advantages that such forms of information technology offer have not been lost to construction companies as more and more organizations are quickly turning to low-cost and easily installable forms of such material tagging to streamline their inventory management systems. As such, automation of material handling is possible, thereby, increasing productivity, utilizing available space better, achieving higher customer satisfaction levels, minimizing operating costs, and boosting material flow. It is not unknown for a system that combines RFID technology, mobile and web portal software to have a significant impact on the efficacy of material and information flow at construction sites. Without doubt, information technology when used in this manner can have a rippling effect resulting in superior supply chain control and management system within a construction organization.

RFID Technology in Materials Management

Radio Frequency Identification (RFID) gained prominence in various industries and commercial establishments by virtue of simple and yet robust design. Today, markets are awash with RFID based card technologies. Despite its popularity, RFID has been rather a slow entrant in the construction sector. In recent years, however, construction companies have begun to increasingly look towards RFID technologies for ensuring robust implementation of asset and facilities management systems. While RFID technology began as a costly route for identifying asset and tracking material movement, it has today grown into a versatile tool that can help construction companies. RFID components have become cheaper and, thereby, tending to be a competitively priced solution in comparison to other forms of material tracking and inventory monitoring systems. The salient feature of RFID technology is that it allows passive or active scanning of the material to which a RFID tag or card is attached. A practical example and commercial success can be said to be the countless instances of retail stores and merchant establishments that use RFID tags of various types to keep track of their merchandise within their stores, transit and inventory points. RFID offers the following advantages:

- Improving productivity.
- Obtaining information on materials in real time.
- Tracking of tasks and inventory processes.
- Improved stock control.
- Improved quality control.
- Significant reduction in associated paperwork.
- Minimizing instances of incorrect product despatches at production centers.
- Alleviate safety and health standards.
- Facilitate in web-enabled customer information systems.

Construction companies can benefit from RFID based systems by improving upon their existing material handling and inventory management systems. This, in turn, can contribute significantly towards bringing down labour costs, improve quality and safety, and help in adhere to project schedules and times. Following are few examples of RFID applications in materials management

- **Inventory tracking:** RFID systems are allowed for near-automatic, real-time, almost fool-proof tracking and inventorying. System is for material handling. The ease with which supply chain management can be managed is rendered far more convenient and manageable by RFID based systems. Issuing of materials, transportation, receipt at site, storage at site, issue for installation and end-consumption can all be mapped through RFID system. At no stage of the process are manual entries any longer required than absolutely necessary. In fact, RFID does away with most cumbersome and repetitive processes that take considerable time not to mention manual labour at the cost of lowering productivity and inflating project schedules and timelines.

- **Expeditious Handling of Materials:** Over the years since its inception, RFID enabled processes have constantly enabled streamlining and expeditious handling of materials. Project planners, engineers, and all relevant stakeholders are almost always constantly aware of material movements and, therefore, are able to make plans and schedules in advance. The more positive effect that is often cited by supporters of RFID is that materials that would have otherwise been underutilized can be tracked and inventoried accordingly.
- **Accuracy in Inventorying:** RFID enables providing information on shipping, receipts, issues, and stocks of materials. This feature is turn to helps in ascertaining location of items on a micro scale and, thereby, obviate possibilities of item mishandling or misplacements. Handheld RFID readers are regularly employed in such situations as these are the important and requisite facet of this technology. Data from RFID handheld readers can provide information on the identity and, when programmed, to also provide locational information so important for accurate tracking of items identified.

There are multiple applications where RFID technology can be used in the construction sector. These include:

In water supply projects, RFID technology can be implemented to address materials management issues that are otherwise difficult to manage and monitor owing to complex operational issues. RFID technology can be used to track tools and equipment at construction job sites. Combined with tool tracking and inventory management system, this can significantly speed up in inventorying processes and track equipment and tools wherever these may be deployed. Automated models for control of materials and materials management can be developed using RFID. A combination of RFID and wireless technologies can help track progress in project construction sites. In industrial projects where tracking of fabricated pipe spools are critical, RFID technology can help streamline the inventory and tracking processes. Precast concrete components along with their historical data can be tracked and recorded using RFID. Using RFID technology, development of a prototype for identification and spatial tracking of structural steel components can be designed and developed at constructions sites.

Mechanics of RFID

The entire structural mechanism and functioning of RFID system is dependent on two components. The first being the tag that is attached to items which need to be tracked and inventoried. The other being the tag reader. RFID tags are basically transponders that have an antenna. Each tag is manufactured with a unique identification number that enables RFID readers to identify and track each tag when scanned. RFID tags can either be passive or active. Passive tags are cheaper while also being of limited range, and require RFID readers to actively seek information from such tags. Active tags, on the other hand, have a wider range and are able to push out information on their identities to any RFID readers that may be in vicinity. Consequently, given their features, active RFID tags require some form of power source and can, therefore, transmit their identifying information over a longer range Figure 1 is a diagram of typical RFID system.

Figure 1. How RFID Works

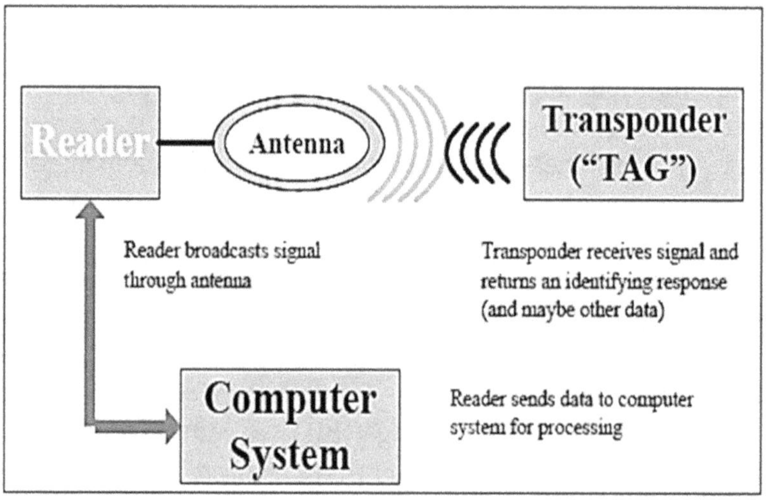

RFID Tags

It is often said that in RFID design, the primary component is the RFID tag itself. RFID tags are the ones that store identifying information which makes it possible for tracking and inventorying processes to take place. RFID tags are of two types, viz., active tags and passive tags. Passive RFID tags do not have any internal power source. They draw their power source from RFID tag reader when brought in proximity of such reader [20]. Passive tags have a shorter range from several centimeters to several feet. Though sporting a shorter range than active tags, passive tags have a longer life span and are considered comparatively robust.

Passive RFID tags contain a microchip and an antenna. These components are mounted using adhesive layers on a substrate to create a tag. A tag may have a paper surface or plastic surface that allows printing of labels identifying the tag and the material to which such is to be affixed. RFID tags can be embedded in plastic cards, in key fobs, in smart labels, or even in packages that are able to tolerate extreme temperature variations or even harsh chemical applications. An example of anatomical structure of passive RFID tag is shown in Figure 2.

Figure 2. Passive RFID Tags

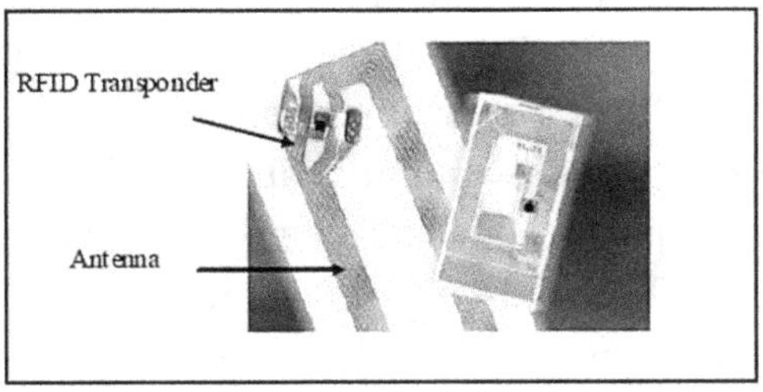

Active RFID tags carry their own internal power source and are often rewritable. These features set active RFID tags apart from their passive counterparts. Consequently, active tags are usually larger, heavier and costlier to produce, procure and maintain. But active RFID tags are also more noise resistant and have a wider read range from several meters to hundreds of meters or more in certain cases. With a relatively short lifespan accorded by internal batteries that can run up to 10 years, active RFID tags usually are deployed in niche environments where passive tags would otherwise be inadequate to the objectives of end-use applications. An example of an active RFID tag is shown in Figure 3.

Active RFID tags are popular in environments that involve water and animal livestock, shipping containers and vehicles, and in situations where long range transmissions are required. Active tags usually have larger memory spaces and, therefore, are able to retain more information. This becomes a very handy feature in case of numerous container shipments over large distances.

Figure 3. Examples of Active RFID Tags

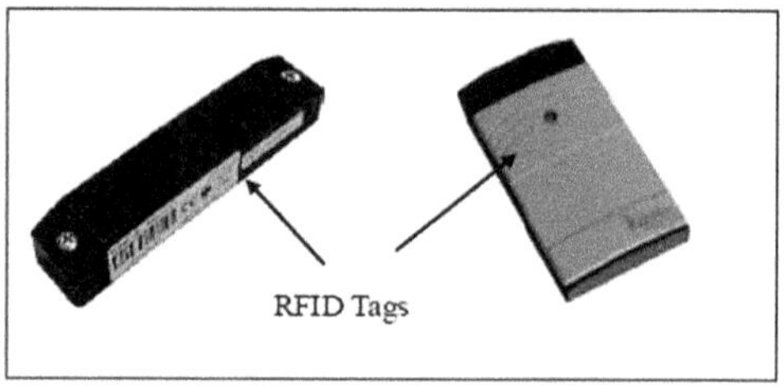

RFID Readers

RFID readers are also referred to as interrogators by some. The reason being that these consist of a radio transmitter and receiver, a control unit, and a type of coupler attached to transponders. The purpose of RFID readers is to interrogate or look for information from RFID tags within a given range. RFID readers may be static or mobile, and are capable of information interchange over short and long distances. Data is exchanged between tags and readers by means of radio waves. While mobile RFID readers can be attached to desktops, laptops, handheld computers, these can also be positioned at fixed locations and at strategic points, e.g., store yards, offices, godowns, etc. making these effective at scanning any RFID tagged materials or items within range.

RFID scanners convert information into digital format for computers to be able to read, process, and report. RFID readers sport features that make them capable of scanning several tags in seconds or less. RFID tag readers such as iCard 3 and iPort 3 can scan 100 tags in a second and identify these individually. Both these types of cards can read RFID tags within a range from several feet to several hundreds of feet in case of iQ tags. Figure 4 shows few examples of RFID readers.

Figure 4. RFID Readers (i-Card 3)

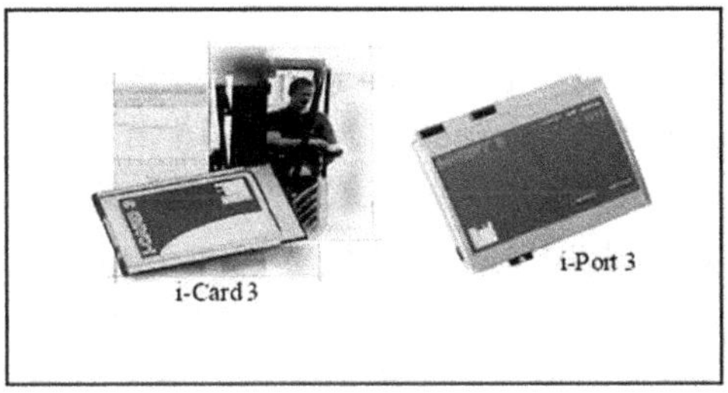

RFID Antenna

An RFID antenna is an element made of conductive materials that enables tag to send and receive data. The purpose of antenna is to create an electromagnetic field within immediate vicinity of the RFID reader. RFID tags which are within range of this electromagnetic field are activated and begin transmitting data. In case of passive RFID tags, this is the only means of communicating information to RFID readers. Active RFID tags on the other hand can transmit information on their own accord by virtue of their internal power sources. Figure 5 shows some examples of different types of antenna used in RFID tags. Elliptically polarized antennae are useful when large number of tags are required to be read at one go, or when such tags are moving at high speeds. Linear polarized antennae are more useful in cases requiring individual selection of tags and in environments where there is restriction in the number of active RFID zons.

Figure 5. Examples of RFID Antennas

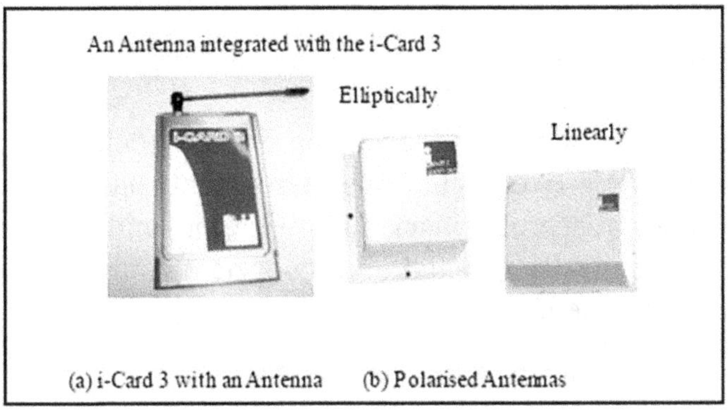

RFID Advantages

RFID technology offers certain advantages which, when looked from the perspective of initial procurement costs and maintenance charges, are considerably attractive than some other technologies that offer similar functionalities. RFID technology allows for automatic identification even in cases where linear sight may be absent. The technology is not only secure, but also allows for a considerable amount of information to be stored and exchanged even in harsh operating environments. Such features make RFID technology an attractive proposition in logistics involving large vehicular movements in hostile terrain and harsh climatic conditions. This technology also makes it possible for cheaper and more effective inventory and material management processes.

The Disadvantages of RFID

RFID technology has certain disadvantages when compared to newer forms of identification systems such as bar code technology. RFID is considerably more expensive and costlier to deploy in supply chain management (SCM) situations. Because RFID uses radio waves to transmit and receive signals, it is vulnerable to interference that may lead to loss in transmission signals; thus, rendering it less than ideally effective.

RFID Based Asset Like Mivan Shuttering Tracking

RFID technology can be applied to construction shuttering materials, e.g., MIVAN. Following are the steps that can be followed for application of this technology to track MIVAN or similar shuttering items.

1. Every MIVAN component / shuttering material / shuttering sheet shall have a RFID tag sticker affixed.
2. RFID stickers shall be attached at appropriate stock yards at site.
3. Each RFID tag affixed shuttering component shall be scanned using RFID reader for preliminary identification and to make relevant entries automatically through an appropriate computerized inventory management system.

4. During construction, each RFID tagged shuttering material shall be passed through a gate that shall have a RFID reader attached. This would enable recording of materials issued from store.

Figure 6. Shuttering Materials Tracking Process

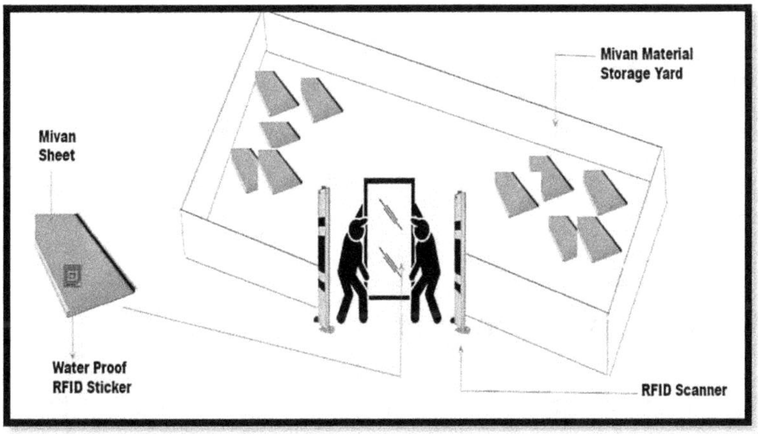

Shuttering material movement inwards and outwards at store can be recorded if the RFID tagged shuttering materials are scanned through a special gate that has RFID reader installed. Appropriate entries can be recorded and passed once RFID tagged shuttering materials are scanned. Figure 7 is an illustration as to how shuttering materials can be scanned by handheld RFID reader.

Figure 7. Handheld RFID Reader

Table 1. Global RFID tag classification

Class 0	Passive	Factory programmable (64 bit only)
Class 1	Passive	WORM with provisions for read/write (96 bit min.)
Class 2	Passive	Short range (4"-18"), R/W multiple, user memory
Class 3	Battery Assisted reader activates, battery powers	Medium range (10'-50'), R/W, user memory, sensors, encryption
Class 4	Active – battery powered	Long range (300'), user memory, sensors…

RFID Applications in Construction

RFID technology by virtue of its feature of enabling uni- or bi-directional communication between tagged objects and RFID readers, makes it a potent alternative in improving productivity, streamlining material movement, and aiding in reduction of labor-related costs and material expenses. There have been a number of publications and research studies on RFID to improve its efficacy and application in construction sector.

RFID has, today, emerged as a tool that can significantly aid in real-time measurement of project information in construction sector. It can be used in assorted areas ranging from equipment management, labour management, safety management, and monitor progress of diverse works involving concrete, pipes, earthworks, structural works, consumption of raw materials and such like.

RFID can be used to track employees and equipment at construction sites. The resultant tracking information collected periodically and/or at short intervals can help in keeping track of cost control. Workers can be issued credit card-sized RFID tags that can also serve as identification badges. Combined with RFID tagged-equipment, a system for tool management and large equipment management can be setup.

Use of RFID tags can help in providing a secure environment within construction sites. RFID tags attached to tools and equipment, and used as identification badges by employees. Such RFID based system can also help ensure that only authorized employees and workers are allowed access to operate complex heavy equipment. Thus, the likelihood of accidents and unauthorized use can be significantly reduced.

Figure 8a. Basic Barcode Structure

Barcode Technology in Materials Management

Barcodes are considered a step up in improving materials management and inventory systems. Bar codes are designed to provide a quick and efficient way for labelling products for improved and accurate identification. The system of barcodes relies on barcode labels and barcode scanners or readers. While the concept in itself is not new, newer barcode symbologies have begun to

Figure 8b. Basic Barcode Access Structure

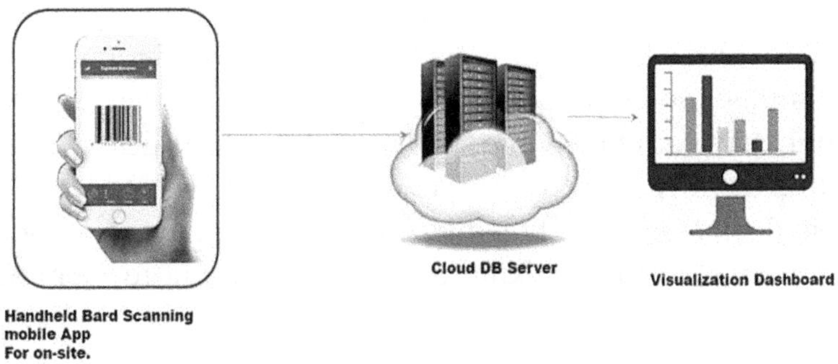

Figure 8c. 1D AND 2D Barcodes

have significant impact in terms of material handling and inventory. Barcodes find a place in almost all industries including construction.

A system of barcodes basically involves a barcode reader, barcode printer, and barcode lables. Several barcode symbolizes can be used depending on the mission and objective. Every barcode symbology has its own unique character encoding rules, error checking procedures, coding, printing and decoding processes. A basic barcode comprises a leading and a trailing quiet zone, a start pattern, followed with one or several characters representing data. As an option, usually one or more check characters along with a stop pattern can also be included to form a complete bar code. An illustration on the anatomical structure of a typical bar code label is given in Figures 8a and 8b.

Barcodes can be either one dimensional (1D) or two-dimensional (2D) type. The former type uses lines of differing widths and spaces to represent data. These, in turn, can be of code 39, code 128, EAN-13, EAN-128, ISBN, etc. Two-dimensional (2D) bar codes on the other hand use matrix and stacked symbols to represent data. These can be QR code PDF417, data matrix, maxi code, etc.

Figure 8c displays various 1D and 2D barcodes. It is said that 1D barcodes were developed with "product identification" in mind. Whereas, 2D barcodes were conceptualized keeping in view the need to accommodate "product descriptions". Limitations of 1D barcode storage, restrict the amount of product information that can be harboured in any given 1D barcode. 2D barcodes offer superior payload embedding, error correction facilities, security of data, and better readability over their 1D counterparts. 1D barcodes while easier to scan and decipher, are hobbled by limitations related to amount of data that can be stored, error checking and readability, not to mention inferior

Figure 9. One Dimension Barcode

UPC-A with Supplemental UPC-E

security features compared to 2D barcodes. The latter are, however, not easily decipherable by the human eye.

The common cause of complaint in case of bar codes is that any video camera can be used to read and decipher the content of barcodes. In such cases, therefore, barcodes are considered improper and inadequate means of storing secure or confidential information.

A barcode reader is an electronic device that scans barcodes printed on labels. The purpose of barcode reader is visually capture the information presented on barcode labels and process the information captured.

Barcodes are broadly classified into three basic categories. These are linear barcodes, stacked barcodes, and matrix barcodes. Compared to the first, the latter two offer significant improvement in terms of storage and damage resistant potential.

Linear Barcode

The most common among all barcode types is the linear barcode. Such barcodes comprise series of parallel lines of differing widths of bars and spaces as depicted in examples shown in Figure 9. Typically, these bars can store a number of characters ranging upwards of 10 to 20 or in certain cases even more.

Figure 10. Stacked Barcode

Stacked

Stacked barcodes are composed of short single linear barcodes that are arranged in stacks as show in the figure below. Being in stacked formation, allows for such barcodes to store a large amount of data in comparison to linear barcodes. As much as 1000 characters can be stored using such barcodes. Notwithstanding such capacity to store data, stacked barcodes are considered not as efficient as matrix barcodes in terms of efficiency in utilizing storage space offered.

Matrix Barcodes (Two-dimensional)

Matrix barcodes were designed and developed in response to consumer demands for more data capacity. Insofar as 2D barcodes are concerned, matrix barcodes are a step up towards offering more information density. The symbology employed in 2D barcodes constitutes a mix of light and dark elements, and comprises circles, squares or hexagons as seen in Figure 11. In construction industry, matrix barcodes are a favoured means of storing records related to equipment identification and maintenance.

Figure 11. Matrix Barcodes

- **The advantages of barcode technology** - Barcode technology is comparatively cheaper than RFID as it involves only paper and ink to produce labels. RFID technology relies on silicon chips, which while are now cheaper than ever before, are still relatively expensive in comparison to paper and ink. Barcodes are scalable, reliable, readily produceable and as accurate as RFID.
- **The disadvantages of barcode technology include** - The chief disadvantage of barcode technology is that barcode scanners need to be in line-of-sight of a barcode label for scanning to be effective. Further,

ordinary barcodes are limited in terms of number of characters that can be stored. Although, this disadvantage is offset by 2D matrix barcodes.

- **Comparison between RFID and Barcode -** A comparison of RFID and barcode technologies is provided in Table 2.

Table 2. A Comparison of RFID and barcode technologies

System	Barcode	RFID
Data Transmission	Optical	Electromagnetic
Typical Data Volume Data Modification	1-100 Bytes Not possible	128-8k Bytes Possible
Position of Data Carrier for Read/Write	Visual contact Not possible	Non line of sight Possible
Reading Distance	Several meters	From centimeters to meters
Access Security	Little	High
Anti-collision	Not possible	Possible

- **Barcodes Applications in Construction** - As explained in the preceding sections in this chapter, a system of barcode significantly simplifies process of identification and data transmission. The construction industry has been witness to proliferation of barcodes in production. Research and practical implementation of barcode systems have established beyond doubt that barcodes can help identify and locate building materials and components at construction sites. These also help reduce the number of instances of loss or improper identification of goods and equipment. With the help of barcodes, it is possible to electronically track movement and location of materials. In certain cases, barcodes have also simplified the process of tracing human resource movements along with materials. These features have made barcodes popular in the construction industry for speeding up production processes and make inventory processes streamlined.

There are instances where construction companies have paired barcodes to time cards or ID cards to gather employee data and details along with cost accounting codes. This approach has made it possible for such construction companies to attribute appropriate productivity data to individual workers. Identification of drawings, documentation and other materials are also possible with the help of barcode system. Despite all the obvious advantages of barcodes, there are certain drawbacks related to range and durability when considered in ground conditions and harsh terrain where remote sites are often setup by construction companies. Barcodes require clarity in printing and visibility at all times. Therefore, smudges, dirt and grime can often compromise the readability of barcodes. It is in such cases that RFID technology holds an advantage over barcodes.

Near Field Communication Technology in Materials Management

Near Field Communication (NFC) is considered one of the latest trending technologies in the field of short-range wireless communications. NFC technology has found relatively widespread acceptance in high-end mobile phones. NFC offers a safe and secure communication bridge between electronic devices. It only requires a mild touch or proximity for NFC communications to kick in between devices. This is often referred to as "tap-in" or "to tap and go" in common parlance. NFC technology requires that devices be brought in proximity of 4 cms. to 20 cms. Though, 4 cms. is considered an effective range for transmissions through NFC to take place.

One of the key advantages of using NFC is its short transmission range that makes it difficult for security breaches to take place during communications. NFC works on a similar principle as RFID and uses magnetic field between electronic devices to initiate communication. Modern mobile phones use NFC for service discovery, sharing of files, and exchanging any other information between two NFC-enabled devices. NFC technology can be found in diverse applications ranging from identification and authentication, ticketing systems in public transports, and contactless transactions at merchant establishments conducted using credit or similarly-equipped debit cards.

Figure 12. Web-based architecture

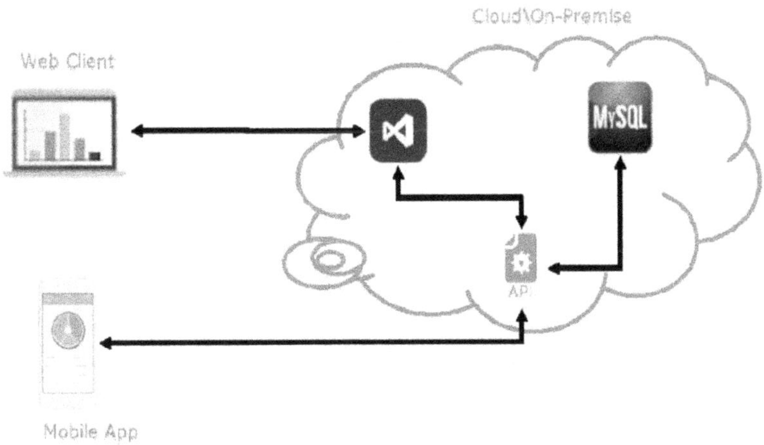

TECHNOLOGY AND ARCHITECTURE FOR MATERIAL INVENTORY SYSTEM

In this section of the chapter, we discuss how a typical web-based architecture can be set up for material inventory system.

Technology Used

- Web Application: ASP.NET, MVC, WCF REST, Entity Framework 6.0, JavaScript, MySQL
- Web Server: Net WCF REST Services hosted on the IIS Server. All the web\mobile app. requests are catered through these services adopting SOA mechanism.
- Security: 256 bits SHA2 Encryption for passwords and sensitive data exchange. Secured (https:\\) data transport using certificates procured and provided by client.
- Mobile Application: Native App (iOS) developed in JAVA, Realm

Components

1. Application & Database Server
2. Web Services APIs
3. Web Client
4. Mobile Application

Figure 13. Authorization technique

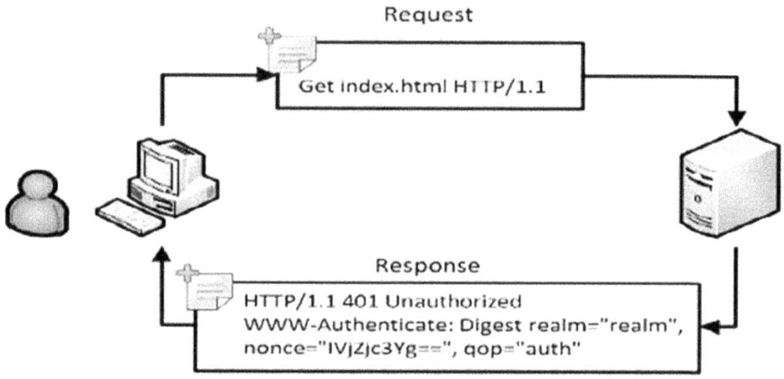

Component Structure

- WCF REST Service, Unit of Work design pattern with generic Repository and Database are components of application server, and ASP.NET MVC 5 is part of client web server. Communication is accomplished using REST WCF service with digest authentication.
- MySQL Database with repository patter used in Entity Framework 5.0 to map database to entities and perform all database related operations.
- WCF REST service is communication point. Any platform application can connect to the Application server with authentication.
- ASP.NET MVC [Model View Controller] is a design pattern/technology used for faster web pages development which is compatible with most of available browsers, though google chrome is our recommended browser.

Mobile Application

Mobile application for scanning assets during audit project. This application communicates with the Application server for exchanging data with DIGEST Authentication process. The WCF rest services created to be consumed by mobile application only. Mobile application also works offline for barcode reading and there is a functionality to sync the offline scanned data with the application server once the device is connected to internet.

Figure 14. Screenshot of windows credential screen

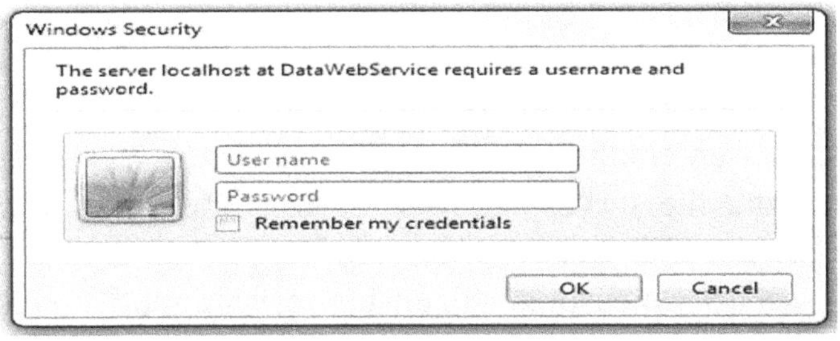

Security

Digest Authentication

Web service repository is core for the application. This services contains all the business rules and data access logic that can be used by web as well as mobile application. User Digest authentication process takes care of application security. Digest communication starts with a client that requests a resource from a web server. If the resource is secured with Digest Authentication, the server will respond with the http status code 401, which resolves as Unauthorized.

Figure 15. HTTP header

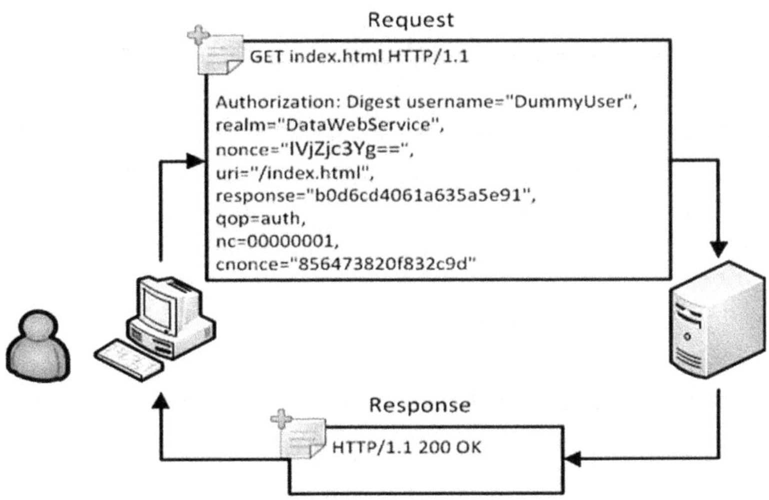

An internet browser responds to this by presenting the user a dialog. In this dialog the user is able to enter his username and password. Note, that this dialog does not show the warning about transmitting the credentials in clear text as with a Basic Authentication secured site.

When the user enters the credentials in this dialog, the browser requests the resource from the server again. This time, the client adds additional information to the HTTP header regarding Digest Authentication.

Following security measures will ensure server application security from external attacks/security threats.

Password Encryption

All User passwords are stored as encrypted format in the database using SHA2 hash with 256 bits encryption mechanism. This ensures extra protection against any malicious activity and hacking.

Data Encryption

Android Phone Data which are downloaded on the android phone is stored in encrypted format and cannot be accessed without authorized login of the android application. Critical asset details like Asset User and Location are not downloaded on the mobile phone. Once the project is completed through App, respective project data is deleted from the phone immediately.

User and Role Based Authentication and Authorization

User and role based access is given to the application and pages of application and android App. Unauthenticated user cannot login to the application or android app. Role base authorization defines the user's accessibility to the functional modules and pages based on rights granted by admin.

Transaction Management

All the transactions at application server are managed by the Unit of work framework.

MOBILE APP CONNECTIVITY AND ACCESSIBILITY OPTIONS

- Using Public IP configuration: Assigning a public IP to the server where application is hosted. Respective URL will be configured in the corporate mobile provided to the user. Mobile app will be able to access the application through WIFI network or mobile web connectivity.
- Using Local IP URL: mobile app can access the local intranet URL (local IP of the machine can be used to access the application within company intranet) through company WIFI intranet. In this case application access or any data upload /download activity wouldn't be possible outside company's WIFI network. However, for this option, all locations where the Application is supposed to be used should be on the same Network.
- Using spread sheet download and upload: We can provide standalone mobile application to the customer where audit project data can be imported, and project data with respective asset scan results can be exported. Exported sheet can then be imported into the Assetrak database via Assetrak local URL.

Hardware and Software Requirements

Hardware Requirements (Server)

- Server Grade machine with Octa Core or above processor (64 bit)
- 16 GB RAM or more
- HDD – 256 GB (SSD hard disk preferable) or more
- Barcode Printer

Software Requirements (Server)

- Operating System: Windows server 2012 Enterprise version or above (64 bit)
- Microsoft Office 2010 or above
- Microsoft Frame 4.5 or above
- IIS 8.0 or above
- Internet browser - recommended Google Chrome
- Browsers should be Java Script enabled

User System Requirements

Laptop/Desktop Configuration

Below is recommended Desktop/Laptop hardware and software configuration for application user to access and use the Web application:

- Core i3 or above processor
- 4 GB RAM or above
- HDD 256 GB or above
- Operating System Windows 2007 or above
- Internet browser - recommended Google Chrome
- Good Internet Connectivity with no firewall restriction/limitations for URLs provided to us for hosting (Hosting addresses) the application.

Mobile Phone Configuration

- **Mobile Hardware:**
  - Quad-Core Processor & above
  - Minimum 1 GB RAM
  - Camera 8 MP or above
  - GPS sensor, Assisted GPS • Wi-Fi and 3G enables
- **Minimum space required**: 500MB. This would vary based on number of assets downloaded for Inventory.

BARCODE BASED INVENTORY SOFTWARE PARAMETERS FOR MATERIALS MANAGEMENT

Following features must be in the inventory software:

General Features

- Asset Creation – Single/Multiple
- Barcode Generation and Printing
- Asset componenting/grouping & Group management
- Label customization - by type, size and information captured
- Storage location maintenance

- Manage user level authorizations
- Email Notifications
- Tracking zero value assets (Assets on Lease or customer assets, etc.)

Asset Audit Features

- Asset tagging updation using app
- Update Inventory note - Damanged, Obsolete, Not in use etc.
- Update user and location info during audit
- Mobile app for audit of assets at third party location
- Asset confirmation feature - User auto-confirms assets with him
- Assets Geo Tagging
- Capture and retrieve asset photo

Asset Audit Reconciliation

- Generate missing, additional, damaged and not in use assets report -Locationwise
- Automated Reconciliation of Asset Audit
- Audit evidence storage and retrieval
- Audit discrepancies emailer to asset owners/users

Asset Allocation Module

- Track Assets by user and owner
- User allocation and unallocation
- User confirmation for asset allocated
- Check unallocated assets (Assets in Spare)
- Unallocated assets aging

Asset Transfer Module

- Asset Checkin/Checkout with workflow management
- Checkin/checkout using mobile app
- Asset transfer form generation
- Transfer assets as Loan, Permanent relocation or to third party
- Aging of assets in transit and asset on loan/sent to third party
- Transfer approval workflow management

Asset Retirement

- Retirement initiation through mobile app
- Workflow management
- Asset photo and document capture at the time of retirement
- Alerts/Notifications on initiation and approval
- Retirement approval workflow management

REFERENCES

Arnold, J. R., & Chapman, S. N. (2004). *Introduction to Materials Management* (5th ed.). Prentice-Hall.

Harris, F., & MacCaffer, R. (2001). *Modern Construction Management.* Blackwell Science.

Kasim & Ern. (2011). The Awareness Of ICT Implementation For Materials Management In Construction Projects. *International Journal of Computer & Communication Technology, 2*(1).

New Strait Times. (2007, May 13). Leverage on ICT in Construction Industry. New Strait Times.

Sarshar, M., & Isikdag, U. (2004). A survey of ICT use in the Turkish construction industry. *Journal of Engineering, Construction and Architectural Management., 11*(4), 238–247. doi:10.1108/09699980410547595

Section 3

Towards Improved Resource Management in the Construction Sector

Chapter 6
Moving Ahead With Enterprise Resource Planning Systems in Civil Construction

ABSTRACT

Commerce and trade around the world require cutting edge tools and business process systems to stay competitive. All around us, this is manifested in how industries—big or small—conduct businesses and adopt new manufacturing and business systems to remain profitable. During the mid-20th century, businesses in the Western world began to adopt material requirements planning to streamline production processes till the finished product. While material requirements planning itself, at that point of time, was not unheard of, the introduction of computers and information technology revolutionised the way material requirements planning was adopted. Researchers and proponents of an integrated information system agree that material requirements planning predated enterprise resource planning (ERP) that has since been a central theme and prime mover of all information and data within organizations.

The concept of ERP is based on the fundamental principle that business processes and information within an organisation can be viewed, analyzed and processed by persons with appropriate authorization at any given point of time. In other words, decisions made and records generated at any point and at any office or remote location of an organization can be viewed almost instantaneously. For organizations, regardless of size and business domain,

DOI: 10.4018/978-1-7998-5291-9.ch006

ERP has turned out to be a facilitator that has transformed how businesses function. At this point, it is of vital importance to note that without Information Technology in close attendance, ERP cannot be implemented. In this, it differs from MRP (material requirements planning) that does not always require Information Technology and associated tools (Shtub, 2001) to function. Notwithstanding this seemingly slender advantage that MRP has to offer, organizations all across the world are keen to adopt ERP in whatever form that best suits their respective needs and business requirements.

ERP brings to the table several key advantages that organizations, especially in key infrastructure sectors, find lucrative to adopt and implement. For example, entire departments that are geographically distant can share data and pool information (Teltumbde, 2000). Branch offices situated at remote locations can access a global form of data at a press of a button. Project planners, engineers, procurement specialists and managers can view data and information that is constantly and continually updated based on inputs received from different offices and sites. This translates into superior business intelligence at all operational levels with critical decisions based on near realtime inputs and actual data. The significance of timing in terms of arriving at business decisions at almost any operational level in businesses cannot be overstated. It is often cogently argued that ERP accords an effective platform (Angerosa, 1999) to act as a universal binding medium through which transactions between customers, suppliers, partnering agents, internal departments and even independent consultants can take place irrespective of geographical distances between them. Indeed, ERP systems have evolved to a stage where enterprises view such as strategic tools that help in gaining critical business advantages and sustainability, rather than being deployed as an advanced accounting and inventory management system. While ERP systems offer a single-point-of-entry and multiple reporting mechanism, the most vital feature is the modularity coupled with secure authentication that make such systems an invaluable asset in the functioning of organizations globally.

The construction industry operates on a disjointed basis more than any other comparable industry anywhere in the world (Markus, 2000). Individual sites where construction activities take place can have their own accounting, management, inventory, design and drawing facilities. Construction companies view their own remote site offices as individual cost centers that are capable of generating and monitoring their own profit and expenses. Yet, it remains crucial for the entire organization, as a whole, that no shred of data or information is left out from the central repository. This allows for other sites to share and pool resources, thereby, reducing individual costs and overheads and increasing profitability and sustainability of organization. The induction and inclusion of an ERP system thus necessitates a central database for all inter-connected, inter-related, or even independent departments and site offices. For a construction organization's management, this central database is a data mine that is vital for making important decisions in real-time.

THE BASIC CONCEPTS OF ERP

Simply put, an Enterprise Resource Planning system is a self-contained environment that links several functional areas of an enterprise such as purchase, finance, accounting, inventory, human resources, production, and logistics. In terms of user interface and underlying features, an ERP system involves one database and a single user interface system per computing platform type. In other words, a user of a Windows-based system will have the same graphical user interface system as would another user on another Windows-based system. However, a mobile device user will have a different user interface that is different from a Windows-based user. In both instances, homologous functions and features have the same origin, i.e., a common central database. This is a significant step up in comparison to yesteryears when different application software had to be used for different functions and departments.

The image below encapsulates the functions and departments that an ERP system can support singly. This embodies the ethos of a single data sharing platform that helps in cutting down costs (Bell, Stukhart, 1987), but more

Figure 1. ERP System

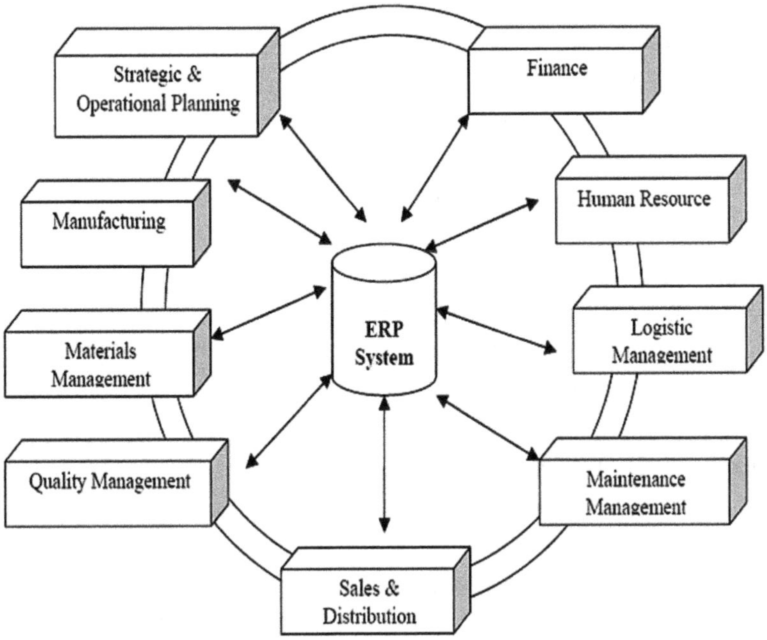

importantly helps in gathering information in real-time or immediately soon after a data entry is made into the system and processed.

In construction companies, ERP can be utilized to perform a range of important functions. These are enumerated as follows:

1. Boost decision making.
2. Bring down costs.
3. Complete projects faster.
4. Improve flexibility in organizational functions.
5. Strengthen partnerships with vendors.
6. Improve inventory control and management systems.
7. Bring about transparency in customer relationships.

The use of ERP within construction industry circles have only been studied by select few researchers. Most of the studies have focussed on processes related to construction and delivery, in addition to facility engineering. Involving organizations with ownership stake and contractual obligations to deliver the projects, such studies have gone on to form a vital bedrock for conducting further analyses on the implications of including ERP within the fold of business operations of construction companies.

As a senior consultant working for a reputed private consulting firm offering ERP solutions once described – ERP systems can and do provide a platform to help propel sales and disseminate project information among managers, engineers, sales representatives, investors, financiers, and customers. In business and residential construction segments, a myriad of activities encompassing land development and acquisition, project development, facility management, warranty and post-handover support, web-enabled services, realt estate sales and configuration management can be handled by ERP systems. The following are the advantages of implementing ERP systems by construction companies:

1. Enabling automation.
2. Integrated Project Management system
3. Flexibility in planning.
4. Responsiveness to global changes and regional mandates.
5. Agile process planning
6. Complete control over life-cycle of projects

ERP MODEL OF INTEGRATED MANAGEMENT INFORMATION

Current business environments and increased competition mandates that flexible reporting tools be available for users at a press of a button. Independence and self-reliance are the keywords. The traditional dependence on any single dedicated IT / MIS department to serve reports on demand to users are on the wane. Time being of the essence, users are now growing accustomed to performing several customer-facing activities through a single application interface. Electronic Data Interchange (EDI) is no longer the preserve of any exclusive or dedicated IT department. ERP systems enable users to vastly reduce paperwork and streamline project planning, through to project management, procurement, sales, delivery, and support related activities. Acknowledgements, order information, invoices, orders, credit notes, etc. can be securely delivered through ERP systems almost seamlessly and with optimal effort involved.

The inclusion or incorporation of electronic mailing facility significantly improves organizational workflow and simplifies processes for the average user. Automating workflow – which by far is viewed as extremely important and beneficial for business operations – can be made possible through the synergy of ERP and email systems.

The most vital component of ERP systems is the database. Without a strong and robust database system that offers data security, active failovers, backup and restore facilities, the advantages that are evident with ERP would not exist. While desktop database systems have grown in stature and sophistication, ERP systems rely on far more complex and considerably more advanced relational database systems that are hosted on dedicated servers running 24x7. ERP systems can be integrated with imaging facilities as well. Real-time or static images can be recorded and collated according to requirements of users. As far as facilities management is concerned, many users find this to be a welcome addition to the otherwise rich repertoire of features that many advanced ERP software solutions provide today. Following is a representative diagram depicting an overview of a modern ERP system.

ENTERPRISE RESOURCE PLANNING SYSTEMS

An ERP system is essentially a closely-knit set of modules or individual software programs that support specific features and functions required to carry out business processes of any organization. A well-designed ERP system is able to support the functions and requirements of multiple departments of an organization. Data from sales and marketing can be analysed and processed by finance and accounting departments. Information produced during manufacturing processes, materials management and inventory handling can be utilized similarly. Human resources and supply chain management along with customer and supplier information can be interwoven. Despite this seemingly complex relationships, each department can yet maintain its own data silo without risk of information contamination arising out of unauthorized and/or unapproved access. The concept of ERP as a system that could boost productivity and enhance management control under a single roof took hold in the corporate world beginning in early 1960s. Initial plans called

Figure 2. ERP Model of Integrated Management Information

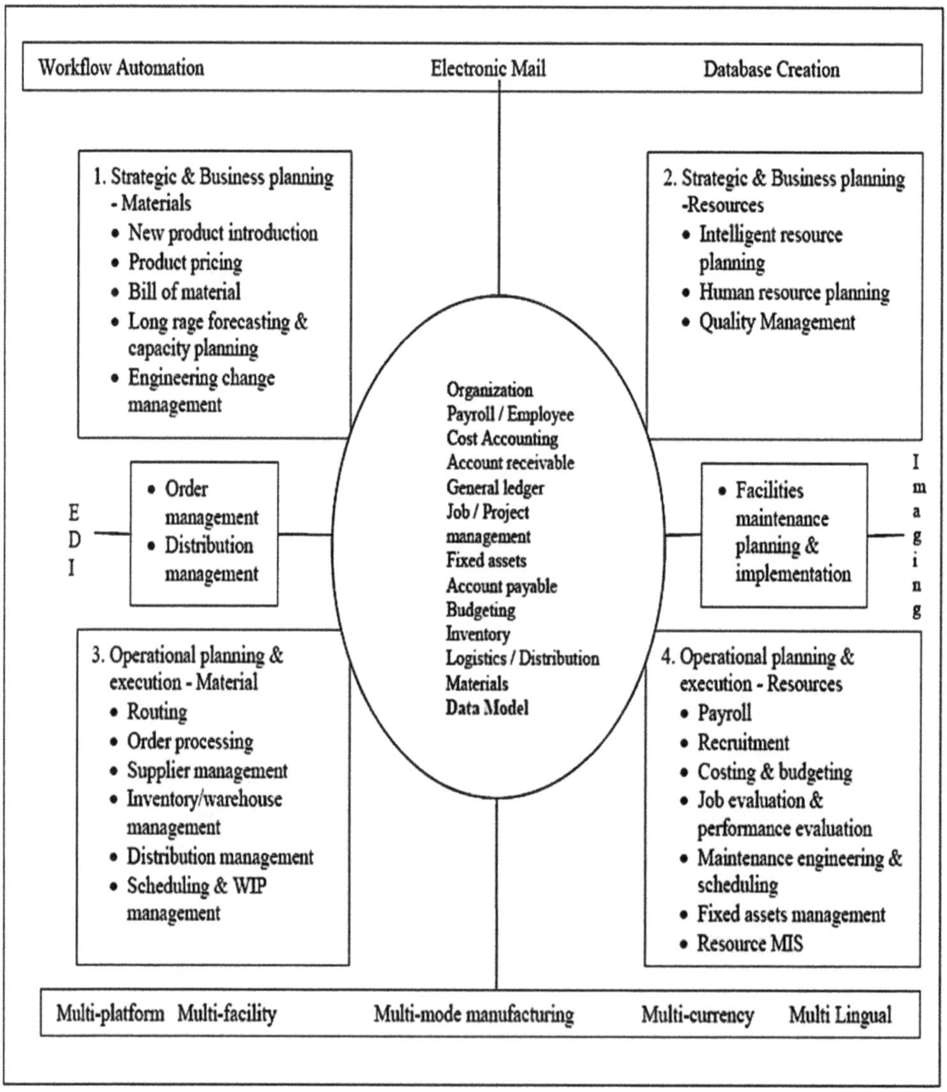

for terms and components such as inventory control packages followed by material requirements planning (MRP) and then succeeded by manufacturing resource planning (MRP II).

Today, ERP systems are bought and sold at rates dictated by license costs, infrastructure expenses for supporting ERP, and implementation fees. Enterprises and large commercial establishments depend on specific ERP system applications that are tailored for specific industries. Such ERP systems are costlier and are sold by established brand names such as SAP, Oracle, People Soft, JD Edwards, etc. Smaller ERP systems such as Microsoft's Navision, QAD, iScala, etc. are popular customers seeking slightly cheaper cost outlay in comparison to more feature-rich and, consequently, more expensive ERP systems sold by companies such as SAP and Oracle.

BRIEF HISTORY OF ERP

1960: Software companies sold application software and systems that was targeted specifically at inventory control. Such systems were designed based on traditional inventory and material handling concepts

1970: During this decade, material requirement planning (MRP) systems were beginning to take hold. These systems were designed based on the concept of the time at which final finished product stage would be reached, and then computed the required components, sub-assemblies and raw materials.

1980: Manufacturing resource planning (MRP II) was the next stage in advancement over MRP. Distribution management services on shop floors was the key focus in designing software applications targeted to streamline material handling and manufacturing processes.

1990: With the advent of microprocessors and latest trends in desktop computing that ushered in a wave of digitalization in unprecedented manner, industries and commercial establishments could afford to invest in information technology. As it became apparent that with globalization new markets began opening up, and sparked competition among companies to stay profitable while cutting costs, organizations began to view merging of departments and all associated business processes under a single system. No longer were departments that dealt in project management, engineering, finance and accounts, human resources, logistics, sales and marketing, and after-sales support left aloof and outside the reach of MRP or its successor the MRP II. Coalesced to form for easing data exchange seamlessly and with least friction, the formulation of Enterprise Resource Planning, or ERP as it is commonly known, took place as 21st century drew closer.

BACKGROUND OF ERP

The term "enterprise resource planning" is used to refer to all activities related directly or indirectly to the functioning of a business. These activities that find place in an ERP system are usually identified with any critical or important aspect of an organization's function. As such these encompass diverse, yet interlinked, areas such as accounting, finance, sales, marketing, production, human resources, inventory, and logistics. Despite the assorted array of ERP software systems that are currently available in the market and in circulation, the basic modules, regardless of design and features, have remained the same. These basic interfaces are as follows:

- **Operations:** This area includes activities such as purchasing, scheduling, quality control, inventory, supply chain management and such other related functions
- **Financials:** This area encompasses most of accounting practices, e.g., accounts receivables, accounts payables, cash flow management, inventory, project management, etc.
- **Human Resources:** Functions that involve labour and staff payments, attendance records, trainings, recruiting, etc. are covered under this area.
- **Customer Relationship Management:** This area includes analytics, sales and marketing, service and support.

TYPES OF ERP SYSTEMS

For a construction company to choose the appropriate ERP system for its use, a select few considerations need to be analysed and discussed prior to inking a deal with an appropriate ERP solution provider or service partner. The considerations that need to be discussed and examined in detail are the different versions of ERP systems that broadly define the scope and applicability involved:

- **Basic Level System:** Such systems only extend basic ERP functionalities. These are rigid and only support a limited number of users and transactions. These systems are also not customizable to a great extent and are often offered on an as-is condition.

- **Middle Level Systems:** These systems support only few basic features. However, these can be customized in future and do offer a greater number of uses and transactional support than basic level systems.
- **Complex Systems:** These systems are more sophisticated than the former two discussed above. Enabling customers to work on any number of modifications with extensive user base and almost limitless transactional support, complex systems have only one downside. These systems rely on latest version of operating systems, application software and database systems all installed on the latest computing hardware available to perform reliably and consistently. While complex systems are a boon for managers, engineers, project planners,

Figure 3. ERP Selection criteria and steps

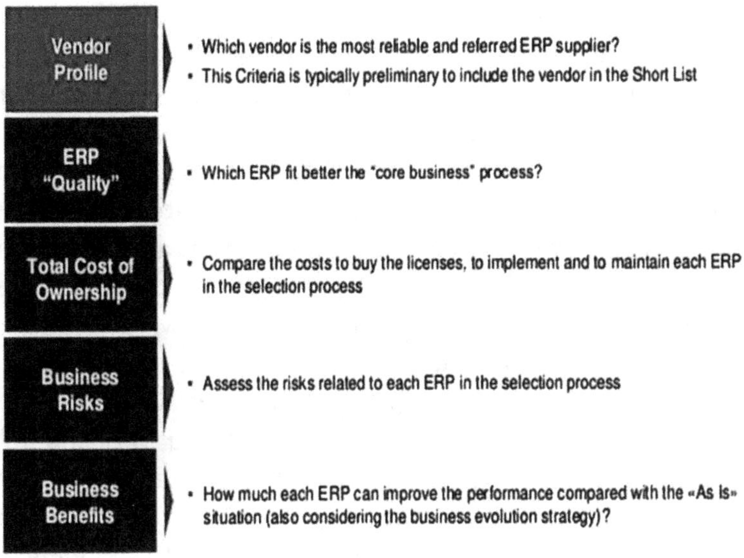

business analysts, business owners, suppliers and customers alike, the biggest challenge is the cost commitment upfront along with annual maintenance expenses required in terms of licensing costs and hardware annual maintenance contracts to be signed with the respective OEMs supplying the computing hardware and associated software required for support complex ERP systems.

A construction company that is seriously on the lookout for an ERP system to manage its operations and stay competitive, the search for a suitable option must begin by analysing market performance of the software vendors. Selection of a suitable vendor should go together with identifying the requirements expected for implementing an ERP solution. Figure 3 is a diagram depicting the selection process involved in choosing an appropriate ERP solution for any organization; but more specifically for construction companies.

ARE ERP SOLUTIONS NECESSARY FOR CONSTRUCTION COMPANIES?

The nature of business of construction companies mandates that organizations in the construction sector gear up to meet strong dynamic market forces. To do so capably, accurately and within the shortest time possible, relevant decisions need to be made based on data available. Data-driven decisions are best made when all relevant information are pooled in and made available for instant processing and analyses. ERP systems are the only ones of a kind that allow such flexibility for organizations that can have sites and branch offices located in far-flung and often remote corners of the globe. While installation of ERP systems are generally costlier and represent a fundamental shift in thinking by policy-makers in a construction company, the advantages offered pose a long-term value and often a gradual offset towards higher degree of competitiveness under adverse market and economic conditions. .

Traditionally, construction companies have been slow on the uptake of information technology and related tools. Paper-based work and information islands created as a result of departments within organization maintaining their individual databases and records, made it extremely difficult for organizations to make important decisions quickly and based on factual evidence gathered from all branches. In the case of construction companies, the situation was all the more glaringly apparent with reports from sites reaching central offices days or even weeks later. Decisions in fluenced by any single department or apex body was often not made after having consulted all available data. The advent of ERP negated these shortcomings and offered a single database that could be updated anywhere over the Internet and accessed at the same time by users with appropriate authorization. Security and connectivity are the hallmarks of almost all modern ERP system. Thus, construction companies can now keep a close tab on all activities taking place at sites and branch offices in near real-time or as soon as data is entered. In cases where automation and IoT has made it possible, data can be fed to the central database with the least amount of human intervention or interaction.

Some of the salient direct advantages of ERP systems are enumerated below:

- Improved cash and fund management
- Streamlining and standardization of business processes and operations
- Optimal utilization of organization's resources
- Optimizing cost of inventory
- Streamlining logistics and movement of raw materials
- Tracking of all operational activities through software dashboard interface
- Closely monitor and track project activity
- Accurate forecasting of project works and funding
- Workflow implementation
- Optimization of human and material resources
- Budgetary allocation and monitoring costs

Figure 4 shows a typical layout of the components of an ERP system. It is to be noted that the important ingredient for any ERP system is the userbase.

Figure 4. The ERP system implementation stakeholders in Company X

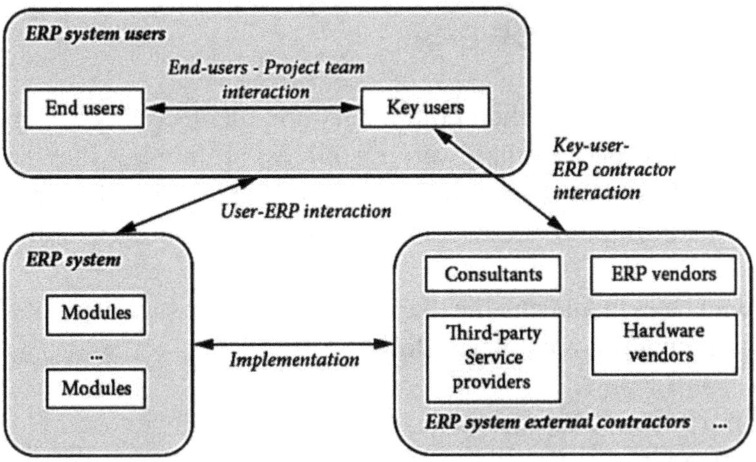

The entire functioning and the underlying planning of an ERP system revolves

Figure 5. Critical Success Factors of ERP System for Construction industry.

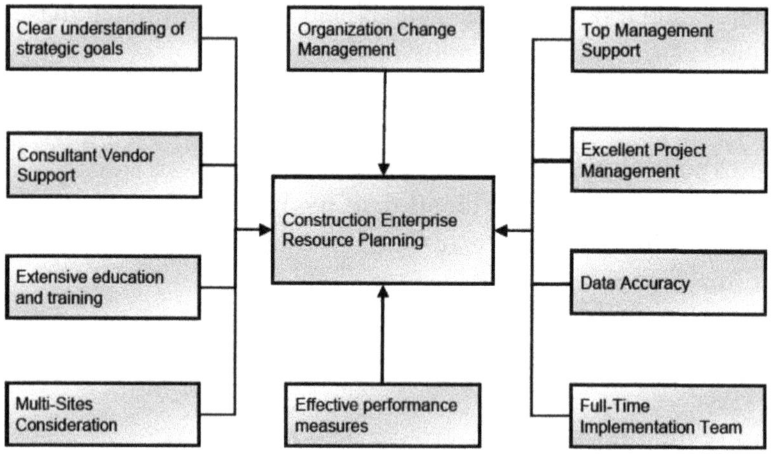

around the users and their requirements. An ERP system would be ill-suited to an organization if the end-users' goals and objectives are neither met nor taken into consideration for implementation at the very outset.

The IT hardware and software components, along with external agencies and parties with whom an organization conducts business with, together form next important layers of any ERP system.

FACTORS THE INFLUENCE SUCCESSFUL IMPLEMENTATION OF ERP

ERP systems by virtue of their complexity and the necessity to put in robust and expensive computing hardware resources, are neither inexpensive nor are guaranteed to work straight out of the box. Several vital factors influence successful ERP implementation. These factors are detailed in a diagram that follows.

As would be evident from the diagram, some of the salient points affecting ERP implementation are noted below:

1. Support from the top management and majority shareholders of organization.
2. A clear overview and understanding of organizational strategic goals.
3. Objective training and education of employees including project managers, planners, engineers, etc.

4. Accuracy of data
5. Dedicated team to carry out implementation.

PROMINENT ERP BRAND NAMES

While respective advocates and supporters would put one before the other, it is widely and generally accepted that the big names in ERP software development and marketing are SAP, Oracle, and Microsoft. In India, Tally is recognized as one of emerging ERP software publishers. Although, all ERP software systems offer a similar degree of functionality that is common in almost all cases, there are certain features and functionalities offered that make these unique in some respects. In the following section, we discuss some of the distinguishing features that are offered by some of the major ERP software publishers.

SAP ERP

SAP is ERP software produced by SAP SE company based in Germany. The initials SAP stand for Systems, Applications and Products in Data Processing. SAP has a worldwide following and is considered to be fastest growing ERP company. The software itself consists of several modules – each of which have been designed to cover almost every aspect of business operations of any organization. SAP supports its own Structured Query Language necessary for fast data transaction operations and has its own programming language in the form of ABAP. Some of the important modules of SAP are as follows:

- SAP Controlling (CO)
- SAP Financial Accounting (FI)
- SAP Sales and Distribution (SD)
- SAP Production Planning (PP)
- SAP Materials Management (MM)
- SAP Quality Management (QM)
- SAP Human Capital Management (HCM)

SAP Financial Accounting (FI)

One of the key modules of SAP that is used for making decisions is SAP FI (Financial Accounting). This module is especially helpful in tracking financial and accounting data spread across multiple organizational entities, currencies, languages and charts of accounts. This module primarily deals in fixed assets, cash journals, inventory, tax accounting, and general ledger, accounts receivable, accounts payable, etc.

SAP Controlling (CO)

This module, as its name suggests, is used for coordinating, monitoring and optimizing processes in an organization.

SAP Sales and Distribution (SD)

This module is one of the vital components of SAP ERP system as it is used for managing billing, shipping, sales, logistics and transportation of products and services in an organization.

SAP Production Planning (PP)

This module is described as ERP Central Component (ECC) that assists in preparing business plans for manufacturing, sales and distribution of goods. It is considered a vital component that facilitates in integration of other ECCs, viz., Materials Management (MM), Plant Maintenance (PM), Quality Management (QM), Human Resources (HR), and Sales and Distribution (SD).

SAP Materials Management (MM)

As the name suggests, this module is essential in managing materials – be it either products or services, or both – in an organization with the objective of improving growth, productivity and lowering costs. The role of materials management is extremely important as it involves ensuring that there are no gaps nor shortage of materials in the supply chain leading up to the point of manufacture and then on to final delivery stage. With SAP MM, procurement processes and material management activities are considerably eased, thus, ensuring that organizations are able to conduct business processes within time and budget.

Structure of SAP Materials Management (MM)

SAP Materials Management consists of several organizational levels. These are classified as follows:

- **Client:** This represents a group or a combination of legal, organization, business and/or administrative units with common interests and goals.
- **Company Code:** This represents an independent accounting unit within an organization. Each company code can have its own associated balance sheet and profit & loss statement.
- **Plant:** This represents an operational unit inside a company code.
- **Storage Location:** This represents a set of data that is specific to a particular storage location. For example, stock levels are examples of data maintained for respective storage locations.
- **Purchasing Organization:** This is an organizational unit that handles procurement of materials and services for one or several plants. It also negotiates with vendors for making purchases. The purchasing organization is legally accountable for all purchase transactions that take place with vendors and third parties.
- **Purchasing Group:** Purchasing organization is divided into individual purchasing groups that are also referred to as 'buyer groups'. These groups are responsible for all daily purchasing activities. A single purchasing group can serve one or several purchasing organizations.

SAP MM Cycle

The SAP MM has a set pattern or cycle consisting of the following:

- Purchase Requisition (PR)
- Purchase Order (PO)
- Goods Receipt (GRN)
- Store Stock.
- Material Issue Slips (GI)
- Inventory Stock

Features of SAP Materials Management

Procurement of organization is an all important process. A sizeable cost of operations is composed of purchases and services of vendors hired. It is not, therefore, without reason that SAP, as most major ERP systems, offers sophisticated materials management module. The major advantages that materials management module of SAP are as follows:

- Generate reports on material consumption.
- Reduce material wastage.
- Reduce manual errors.
- Simplify procurement process.
- Speed up procurement operations.

There are, however, certain disadvantages. For smaller organizations, implementation of this SAP module alone can pose significant problems. These are as follows:

- The initial cost for procurement, installation and commissioning of computer hardware and associated systems necessary for supporting ERP software can be quite high.
- ERP software systems by virtue of the features that are combined, require a certain degree of technical competence and skills. Training and functional support are long-term investments that are to be borne in mind.

These drawbacks are significant stumbling blocks for small construction companies with limited budgets.

Oracle ERP

Oracle is a global leader in database systems. It, therefore, is not surprising that Oracle offers Oracle ERP Cloud and Oracle Apps as stellar examples of ERP products usable for organizations of all sizes and in almost all sectors. Similar to SAP, Oracle ERP products offer several modules to support functions and operations critical for survival and running of day to day business activities of organizations. As such, Oracle integrates assorted and essential modules, viz., accounts, finance, inventory, assets, project management, procurement, human resources, etc. in its ERP software solutions.

The legendary reliability and focus-driven approach of Oracle ERP products have been supported in core by Oracle's relational database systems (RDBMS). Oracle's RDBMS support storing and retrieval of humongous volumes of multimedia data, and extend features that include advanced form and report generators, object-oriented visual interface script language, SQL scripts, stored procedures, cross-platform support, and support for mobile and web applications. What makes Oracle expensive and yet the most reliable alternative in IT domain is its time-tested ability to harbor fail-safe mechanisms and built-in data protection measures that allow huge volumes of data to be stored safely and securely.

Microsoft Dynamics

Another prominent player in the ERP software scene is Microsoft's Dynamics. It addresses key corporate functions of manufacturing, finances, human resources, and supply chain management. Similar to Microsoft's 365 office productivity platform, Microsoft Dynamics offers cloud services and extends features such as artificial intelligence, mixed reality and marketing. For most organizations seeking core business process management, these appear to be more than necessary. However, for companies that are seeking to leverage the advantage of having offices and sites across diverse geographical locations, Microsoft Dynamics is a viable option that promises to offer seamless integration and adequate performance comparable to the best in the market.

Modules in ERP for Construction Companies

Construction companies are no different from any other manufacturing industry. However, there is a distinguishing feature that many argue as being a point of separation where requirements for an ERP system for construction companies diverge from those of any other sector. By virtue of business requirements, almost all construction companies have offices at site locations that are geographically apart and situated in remote regions. This, therefore, requires that any ERP system should be designed and implemented in a manner to account for business intelligence that requires factoring in resources and

Figure 6. Critical Success Factors of ERP System for Construction industry.

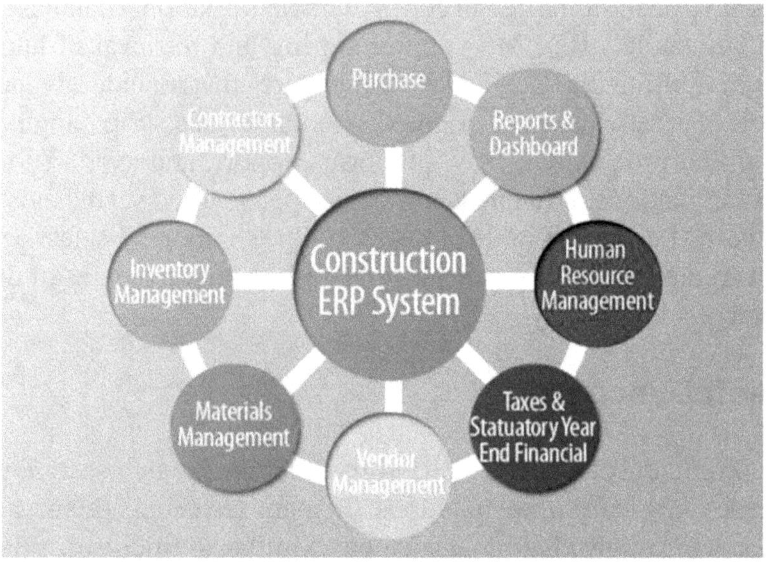

time scales required for delivering goods, materials, and personnel at remote locations.

While the construction sector in India is not renowned for adopting newer IT technologies as and when these appear, ERP systems and related technologies are invaluable for maintaining critical competitive edge. Figure 6 illustrates the various business processes and functions that an ERP system tailored for construction sector ought to support.

REFERENCES

Angerosa. (1999, Oct.). The future looks bright for ERP. *APICS - The Performance Advantage*, 5–6.

Bell, L. C., & Stukhart, G. (1987). Cost and Benefits of Material management Systems. *Journal of Construction Engineering and Management*, *113*(2), 222–234. doi:10.1061/(ASCE)0733-9364(1987)113:2(222)

Lynne Markus, M. (2000). Learning from adopters' experiences with ERP: Problems encountered and success achieved. *Journal of Information Technology*, *15*, 245–265.

Shtub, A. (2001). A framework for teaching and training in the ERP era. *International Journal of Production Research*, *39*(3), 567–576. doi:10.1080/00207540010009714

Teltumbde, A. (2000). A framework for evaluating ERP projects. *Journal of Information Technology*, *38*(17), 4507–4520.

Chapter 7
Cloud Computing and Its Implications for Construction IT

ABSTRACT

By nature, a construction company needs to have data and information readily available at any stage and at any location. Given the often unstructured and fluid nature of construction activities, construction personnel face the challenge of collecting as well as collating data while on the move and at work under field conditions. With the advent of cloud computing and the rapidly increasing reach of the internet through wireless and mobile network carriers, construction companies are increasingly beginning to explore the power of information technology through cloud and IoT platforms. While at one stage, the cost of maintaining IT infrastructure was expensive for construction companies and engineering firms of all sizes, the rise of cloud computing has enabled to rationalize costs and investments while retaining high levels of productivity and service. This chapter discusses the various options that construction companies now have to choose from to streamline operations and increase productivity.

A survey of operations of any construction company is bound to reveal an important facet. That is the distributed nature of workings and revenue generating centers that are distributed and generally located far from the nerve center itself. It is often argued, and perhaps quite cogently, that a

DOI: 10.4018/978-1-7998-5291-9.ch007

construction company survives by numbers. Number of projects that it has, the number of billing cycles that are produced per construction site, the number of employees involved, the number of suppliers involved, etc. The construction sector is purely a numbers game. Beginning with engineering, accounting and right through to human relations management, supply chain management and customer relationship management, an organization in construction sector relies on the numbers involved. It is, therefore, evident that collaboration and integration of data, information and resources are extremely important for a construction company to survive and in the final analysis stay ahead of the competition. It is essential that all departments must collaborate in order to improve efficiency, increase productivity and create an environment where resources are utilized to the maximum possible extent so that construction projects can be successfully executed and stated objectives achieved in a timely fashion. These stated objectives involve metrics such as completion of project schedules within the stipulated time and budgetary allocation of costs, and ensuring quality and safety at all times throughout design, construction, implementation and handing-off phases. For the past few decades, the construction sector in India has witnessed increasing efforts to implement new techniques, and introduce cutting edge technologies and tools to assist in production efficiency and to bring down costs and lower duration required to execute projects successfully. During all this while, the construction sector has been cognizant of the fact that at no stage should production quality, efficiency and safety suffer in the quest for increased productivity and lowering of costs and time. The dynamics of construction industry demand that the construction supply chain itself should yield consistently high value for money in addition to quality of workmanship and optimized construction cycles. All of these must be aided and supported by data and information that is up to date at all times.

The nature of a construction company mandates that mobility and availability of information be held paramount. This is all the more important given the unstructured and utterly fluid nature of construction activities that take place. Workplace difficulties and challenges combined with hazards that construction personnel face daily has, therefore, been a fertile ground for adoption of cloud computing technology (Purohit, Jaiswal, Pandey, 2012). Cloud computing is an utterly novel form of dealing with data and information using computers and mobile handheld computing devices. Cloud computing is an amalgam of modern information technologies such as virtualization and the Internet. With the help of cloud computing, businesses, organizations and individuals alike can harness the power of computers and use storage device facilities at the fraction of the cost it would have taken to purchase, own and maintain the required equipment at own premises (MalizaSalleh, Teoh, Say Chan, 2012). Today, cloud computing is available as several bundled options such as Software-as-a-Service (SaaS), Platform-as-a-Service (PaaS), and Infrastructure-as-a-Service (IaaS). There are several other cloud deployment models as these are known. Each of the others are either derivatives or combinations of these core models. This chapter is dedicated to understanding the concepts and the advantages that cloud computing brings to the fore for construction companies, especially in India. Traditionally, construction companies have used software programs that have run off of servers and storage devices installed at local or within the premises of respective organizations. This model is referred to as "on-premise" and it differs significantly in that the data remains localized and contained in silos that are considered relatively free from access in the outside world or the Internet. However, with the advent of cloud computing, such on-premise deployments have begun to be relegated to a point where the latter forms are considered expensive in comparison to the "elastic" and on-demand scaling of resources quickly and cheaply. For example, an organization can rent a service requiring four core or eight core processors on cloud without actually venturing into the market to buy a costly physical machine, e.g., servers, and then not use it beyond a certain short period of time. Thus, cloud computing progressively has simplified how the world views computing and managing computing resources in cheap, efficient and secure manner. For construction companies, this has meant that their employees can access data and upload information from any part of the globe regardless of how remote locations can be, at the press of few buttons via the Internet. No longer do construction companies need to worry about maintaining costly IT infrastructure and meet regular annual renewals that can run into several hundreds of thousands of Indian Rupees in certain cases. Aided by cloud computing technology, users can manage and handle data encompassing project management, ERP, human resource information, supply chain management data, etc.

While cloud computing has made deep inroads into commercial activities and in many industrial sectors, for construction companies it has been shown to promise a lot and deliver the same in relatively short span of time. While larger construction companies are still quite a way off from moving off from on-premise IT systems to cloud computing environment, for smaller and many mid-size organizations in this sector cloud computing offers a radical, workable, secure and cheaper solution overall.

CLOUD COMPUTING AND THE CONSTRUCTION INDUSTRY

By virtue of having the Internet as its backbone, cloud computing has attracted employees of construction companies. Employees who require almost constant access to vital business critical information and data regardless of their own physical location, to make decisions that can ultimately affect an organization's profitability, have grown to regard cloud computing as a serious and dominant information technology. Central and branch office employees of construction sector companies benefit from cloud computing's advantage to serve data and information that can be accessed almost instantaneously as these are uploaded and/or processed by employees working at sites and remote locations. No longer do offices have to rely on individuals or teams tasked with assignments to go out to remote locations to bring in data. Thus, cloud computing almost has obviated the concept of offline data and information source, and has ushered in a new era of an online and an almost seamless experience in accessing and collating information as and when data is uploaded and is available.

While traditional client-server hardware and software combinations do offer similar advantages, cloud computing significantly broadens the scope and widens the reach from where data can be harvested and pooled in securely and almost instantaneously. Thus, it can be said that back-office operations that would have otherwise necessitated greater involvement of manhours and precious resources to produce budgets, expense sheets, cash flow reports, project schedules, etc., stand to gain and be liberated from such shackles of costs and time through adoption and use of cloud computing and allied technologies.

While cloud computing cannot replace actual meetings with clients at sites and remote locations where projects are in operation, it can help simplify data processing in areas such as billing, logistics and supply chain management, inventory management, human resources and payrolls, accounts and cash flow, etc. Not only is cloud computing beneficial to employees at remote locations and sites, employees in city offices, too, are able to view and process information as soon as the data is uploaded from any remote location or connected office or site.

One of the chief advantages of cloud computing does not rest solely on harbouring data and information that can be accessed securely from any part of the globe connected to the Internet. Although, that is regarded as one of the top features of cloud computing and considering the fact that there is negligible to no risk at all insofar as loss of any data is concerned, the significant factor lies in that that cloud computing enables access to applications as well. No longer do staff have to rush back to their desks or designated workstations in order to work on applications that are either installed or can be accessed in a localized networked environment.

For entrepreneurs and stakeholders, it must always be borne in mind that such path-breaking trends and technologies such as cloud computing must always be aligned with an organization's business processes and needs in order to achieve sustained growth. Construction companies can stand to gain if cloud computing technology is put in proper perspective in the scheme of operational interests and functions.

Like any other sophisticated tool or technological approach to construction engineering, cloud computing systems should be amalgamated and taken into consideration well in advance in designing and laying out strategies before project work is commissioned and initiated. An early adoption to cloud computing way can doubtlessly simplify business processes and operations later on, and improve the chances of reaping the advantages that cloud computing has to offer.

BENEFITS FOR CONSTRUCTION COMPANIES WHEN MOVING TO CLOUD

What Is Meant by Moving to the Cloud

Cloud computing offers a host of opportunities and significantly redefines the way consumers and businesses utilize and harness computers and computing resources. Traditional applications such as corporate email systems and Enterprise Resource Planning solutions that are reliant on on-premise deployments can now be moved on to cloud computing platforms. So, too, can Building Information Modelling (BIM) systems and Project Management applications be migrated and designed suitably for operations via cloud computing platforms. One of the leading cloud-based solutions that appeared in the market as collaboration portals was Union Square. This and 4Projects allowed users to share drawings and operate collaborative digital spaces to exchange information centralized around given construction projects. Needless to mention, of course, this automatically translated into early savings as reproduction costs and printing costs associated with drawings that required modifications as and when needed, were minimized.

Customer Relationship Management solutions such as Project Sales Achiever and Microsoft CRM relied on cloud computing deployments to present consolidated views of an organization's business. It was not long before e-tendering and e-procurement began to proliferate bringing with them the resultant effect of reducing reliance on paper-based communication trails such as orders, invoices, delivery and consignment notes, etc. A new industry standard known as "CITE" or Construction Industry Trading Electronically was formed. This allowed construction companies to tie up with partners such as contractors, consultants, suppliers and merchants and form a supply chain management that produced little or no paper waste at all. The end result was lowering of costs and simplifying communications thereby enhancing response times and helping in improving overall efficiency of business processes and operations.

CORPORATE MAILING SOLUTION ON CLOUD PLATFORMS

Cloud Computing can help save a lot of money when it comes to setting up an IT infrastructure. At costs that are considerably lower than that required in case of on-premise deployments, businesses and individuals can speed up individual start-ups with ease. Further, maintenance and running expenses in case of on-premise deployments can be costlier than that of cloud computing deployments. Cloud computing also offers the added advantage of comparable, and in certain cases even better, speeds in setting up and implementation of systems and solutions at similar or lower price points.

The Advantages of Corporate Mailing Solution Setup on Cloud

For businesses looking to setup corporate mailing solutions, cloud computing can be relatively painless exercise and cost-effective at the same time. A cloud computing service provider needs to be chosen after comparing various options and offerings available in the market. A cloud computing service provider (CSP) is responsible for setting up distributed systems and hardware to manage and store customer data. While this may seem like setting up an on-premise deployment at first, CSPs can and do often rely on several data centers located at various locations around the world to service their clients. The advantage of having several inter-connected data centers can be viewed as two-fold. First, at no point are CSPs threatened with consequent fallout and damages should one data center fail. Applications and data are replicated throughout CSPs own respective network of data centers with backup clusters and failover systems in place to route their clients' requirements to alternate working data centers should any one fail. Second, with several data centers located at several locations around the globe, CSPs can rationalize costs and lower their own operating expenses by having multiple clients share computing hardware resources. Cloud computing maximizes computing hardware and resources thereby ensuring that idle time is reduced. Consequently, investment and maintenance costs are apportioned across a wide spectrum of clients for CSPs and which, in turn, leads to operational profits. For customers of CSPs, all these translate to superior speed and ease of use as long as availability of Internet is guaranteed. With falling rates of Internet, cloud computing is now appearing to be increasingly lucrative for construction companies of all scales and sizes.

Currently, vendors such as Google, Microsoft and IBM all offer cloud-based mailing solutions that also come packaged with active mail spam filtering and antivirus services, archiving options, encryption features and many other tools that are offered to improve end-user experience while ensuring that the tasks of mail administrators are not unduly complicated. In certain cases, cloud mailing solutions are often layered with customer relationship management (CRM) systems. Such systems can be used to speed up business processes and facilitate in a greater degree of control with partnering vendors and customers. Thus, the advantages that cloud computing offers in terms of mailing solutions and supported by mail storage and archiving can appear attractive for construction companies looking to modernize and, simultaneously, lower operational costs associated with maintaining corporate mailing systems.

The following list broadly encapsulates the advantages that cloud computing has to offer for businesses and entrepreneurs invested in the construction sector:

Lowering of initial investment costs by shifting from capital expenditure model to an outright operating expenditure model. This shift from CAPEX to OPEX model in entirety can translate to significant savings insofar as setting up of infrastructure, buildings and support systems are concerned.

Considerably lower and predictable cost of ownership can be had instead of costlier on-premise deployments that may require periodic upgrades and expensive renewal plans.

Management of IT services are also much easier as that entails enabling / disabling services through software dashboards that are in many cases offered by cloud service providers to customers seeking greater degree of control and autonomy from CSPs.

IT personnel who would have been engaged in setting up and regularly maintaining on-premise setup can be engaged in other tasks that can help them achieve greater productivity.

The very "elastic" nature of cloud computing – meaning that services can be rented or opted out from within minutes with the help of associated CSP – allows for more efficient use of available funds. For construction companies in India, for whom cash flow has often been a problem, this can be useful in cases where sophisticated features and options can be let go while retaining only basic functionality.

Table 1. Comparison between traditional and cloud computing deployments

		Corporate Mailing Setup Involving Traditional On-premise Infrastructure	Corporate Mailing Setup on Cloud Computing Platforms
Delivery Model	Responsibility	Deployment and maintenance of IT systems and associated infrastructure are to be conducted using internal resources. You deploy and maintain the infra and application using your internal resources	Cloud Service Providers (CSPs) provide fully managed services. Typically contracts with CSPs are entered into by signing service level agreements (SLAs) that guarantee a certain level of service and offer to penalize CSPs in case service quality is affected beyond permissible levels.
	On-boarding	On-boarding process can take up a while with equipment procured, tested, deployed, fine-tuned and managed, subsequently.	The process of on-boarding is greatly simplified as CSPs take care of almost all IT infrastructure setup and activities. Services can be setup and initiated usually within a day or less with systems up and running at operational readiness.
Infrastructure	Responsibility	Responsibility to purchase, test, deploy, provision and maintain rests with businesses.	With the help of cloud computing, any or all associated costs related to on-premise deployments are left out. Businesses are only required to pay monthly or annual fees to enjoy the conveniences of corporate mailing plans without having to invest any time or effort to setup IT systems.
	Data center	For most construction companies and other organizations in comparable industries, setting up of data centers is neither a feasible option nor a business goal. Data Centers, to be judged efficient and reliable, need to have certain level of certified reliability as required by governing bodies associated with the IT industry. Those objectives are not only out of line for construction organizations but can also draw away much needed manpower and resources to deploy and maintain. In an age where optimization of costs can often mean survival in an increasingly competitive market, on-premise deployments may often look like investments meant to be bypassed for construction companies.	Cloud computing service providers such as Amazon, Google and Microsoft offer standardized high-grade cloud offerings. Such usually have multiple available zones, backup and redundancy systems, and failover systems. Services offered by such CSPs are often certified by authorized certifying authorities in the form of CERT, CMM and ISO.
	Provisioning	Provisioning IT resources is an activity that requires setting up of hardware and supporting infrastructure for activities that can take place in future. Provisioning can be expensive as there is no guarantee that such IT systems and infrastructure once setup, can be set aside for other tasks or sold off at acceptable rates.	Most CSPs do not require fixed contracts beyond calendar year. In some cases, cloud services can be terminated within a day. With no upfront fees required for provisioning, customers and businesses need only pay for the resources that are being used. Fees are payable usually on a monthly basis. In case resources are required or needed to be released when no longer needed, CSPs can help to do so without any considerable degree of time lag or costs. Only rents or usage fees need be paid to the extent cloud computing resources are used.
	Technology refresh	With on-premise deployments, technology once purchased is like a lock-in investment. Usually IT investments are considered relevant for a period no more than 5 years. Upon modernization, such assets are considered a liability with limited or no use at all and, as such, rendering those as scrap or only as useful as spares to be scavenged for other similar products and systems if still in use.	One of the key advantages of opting for cloud computing is that CSPs always ensure that they have the latest and modern IT hardware and systems in operation. Aside from the fact that that is in their own interests for CSPs, for consumers that translate to access to latest computing hardware and platforms while paying minimum of costs paid in terms of usage and contracted rents.
	Scalability	Scaling often mandates a rethink of existing IT infrastructure and setup. While data loss is certainly a point to consider in such eventuality, greater concern lies in arranging necessary downtimes and consequent loss of productive manhours and cessation of business operations till such time scaling activities are carried out.	With cloud computing, managing resources and IT systems is greatly simplified. CSPs take care of all the underlying details involving management of actual IT hardware and associated infrastructure. For consumers of cloud computing, that means significant amount of savings in terms of upfront costs, technical manpower, and business operational hours.

continued on following page

Table 1. Continued

		Corporate Mailing Setup Involving Traditional On-premise Infrastructure	Corporate Mailing Setup on Cloud Computing Platforms
Reliability/ Availability	Responsibility	Availability and reliability are critical factors for every IT manager in service in a construction company. Given that vital operations and critical business decisions are reliant on constant availability and trouble-free performance of IT systems, this can pose a significant burden on technical personnel.	CSPs offer disaster recovery options to ensure that business operations of consumers remain unaffected regardless of whether a concerned CSP suffers partial or whole loss of a data center. As such, most CSPs guarantee and claim with utter conviction an uptime of 99.99999%.
	Disaster recovery and Uptime	Disaster Recovery option is a vital part of setting up of data center operations. For organizations, this typically implies setting up of a similar data center in a different geographical location with similar features and capabilities as that of primary data center. It, therefore, is no surprise that costs and investments are automatically doubled to ensure that a DR site is up and running to achieve a goal of 99.9% uptime to support ceaseless business operations and high degree of reliability. At a time where market forces mandate lowering of costs wherever possible to achieve an optimum operational balance of productivity vis-à-vis costs, setting up of a DR may appear to be counter-productive and significantly cost-prohibitive in the long-run.	CSPs offer disaster recovery options as a feature that is built-in with almost all plans sold to customers. CSPs over-provision their systems so as to ensure that customers can work 24x7 unaffected even in case of severe malfunction at any of the CSPs' data centers. The promise of assuring 99.99999% uptime thus comes as no surprise as customers' operational requirements are seamlessly transferred to alternate disaster recovery sites within the framework or lattice of CSPs' data centers.
	Data Safety and Durability	Data durability is a term that implies a safe and reliable feature to protect data that is stored. Data durability is a term that can be said to be the other face of the coin that has disaster recovery option stamped on it. In other words, data durability and disaster recovery options go hand in hand.	Cloud computing offers a high degree of data durability at significantly lower costs as compared to traditional on-premise investments and on-premise DR setups.
	Monitoring and NOC	One of the key maintenance issues involved in setting up and maintaining data centers is the deployment of automation tools to monitor service and infrastructure uptimes. Teams comprising trained IT personnel are required to work round the clock to ensure that all systems work without problems.	In case of cloud computing environments, digital systems and CSP allocated 24x7 network operations center (NOC) teams ensure that all systems function smoothly. As such, for construction companies this translates to significant savings in manpower and material costs, not to mention savings achieved in terms of manhours.
Scalability	Responsibility	Scalability is a term that is associated with the amount of IT hardware and allied resources that need to be made available to support business operations. For construction companies with core focus in the business of construction alone, this implies taking on added responsibility of setting up and maintaining data centers on a scale that is neither viable nor profitable in the long run.	In the case of cloud computing, businesses, regardless of their respective spheres of operations, need not be concerned with the details of setting up reliable IT infrastructure. For construction companies that translates to significant amount of savings as vital manhours can be invested and channelled into areas that are more focused in terms of business objectives.
	Scale up/out	Scaling up or down involves long and expensive purchase cycles and cost-intensive capital expenditure.	For consumers of cloud computing, scaling upwards or downwards usually involves selecting relevant options that are usually intuitively placed on web-enabled software dashboards made available by respective CSPs.

continued on following page

Table 1. Continued

		Corporate Mailing Setup Involving Traditional On-premise Infrastructure	Corporate Mailing Setup on Cloud Computing Platforms
Security	Responsibility	With on-premise deployments, the aspect of security needs to be looked into and carefully planned and managed at every stage by organizations on their own.	With cloud computing, corporate emailing infrastructure usage policies can be easily configured and setup using dashboards provided by CSPs.
	Infrastructure security	Physical security of entire infrastructure has to be managed and monitored by organizations.	In case of cloud deployments, the responsibility of protecting infrastructure lies with the respective CSPs.
	Data Security	Data has to be stored encrypted with controlled limited access. You need to design tis into your info sec working.	CSPs provide built-in features that cover data access control, mail usage policies and other relevant security measures.
	Email Security	Organizations need to plan and procure a gateway mailing solution (which is also referred to as scrubbing) to remove unwanted mails or mailware.	Almost all top of the line cloud mailing services offer integrated email gateway solutions that can detect and scrub most malware. In all cases, high performance is assured with the help of technological backing of world's most influential software companies such as Google and Microsoft.
	Network Security	The responsibility of designing and deploying WAFs, intrusion detection systems, TLS, attack mitigation systems, etc. lies with organization. This is a vital aspect as in the absence of such systems, business critical data may be lost, and emails compromised.	In the case of cloud computing, resources are secured at multiple layers with IAM, authorisation, VPCs, WAFs, firewalls, TLS, etc.
	Data Ownership	Data is stored in an environment that is maintained by an organization.	Although data in cloud is synchronized and backup copies kept at multiple geographical locations for quicker retrieval in case of any incidents, data security and ownership are closely monitored and rigorously guarded by CSPs using latest tools. CSPs neither own data, nor are privy to any information that an organization may upload or download at will at any time during or expiry of a contract.
Application Refresh	Responsibility	Application refresh and updates are responsibilities that rest solely on organizations.	In case of cloud computing, application refresh and updates are carried out by CSPs automatically as part of contractual obligation. Consumers are required to only use applications as and when latest releases are made available by CSPs.
	Updates/ Upgrades	Since upgrades and updates in on-premise environments typically involve downtimes, most organizations prefer to stick to current versions that grow older and finally rendered obsolete through passage of time. As a result, organizations stand to lose on efficiency and usefulness, not to mention suffer security compromises, due to reluctance in upgrading or updating IT systems to the recommended levels as mandated by original equipment manufacturers (OEMs).	In the case of cloud offerings, CSPs typically ensure that the latest updates and upgrades are rolled out periodically without any downtime involved. For organizations that translates to significant savings in terms of manhours, and helps in business functions to be carried out 24x7 without IT systems induced interruptions on account of even routine maintenance activities.

Comparison Between Traditional and Cloud Computing Deployments

The following table illustrates how cloud computing can offer significant advantages over traditional on-premise deployments. Keeping in mind that most corporate IT solutions require costly capital investments requiring purchase of computing hardware, licensed software, setting up of buildings and rooms with dedicated firefighting equipment, captive power generating systems and power backup facilities, water and support personnel, cloud computing offers a far more simpler and cost effective approach at fraction of the cost involved in case of traditional on-premise IT deployments.

Cloud Hosting Benefits That Are More Than Just Savings in Terms of Cost

While moving to cloud-based mailing solution may appear to be cost-effective and cheap in both short and long terms, that is not the only benefit that is to be had by opting for such. Some of the relevant key benefits are listed as follows.

Elastic Provisioning of New Users

Cloud computing vendors such as Google and Microsoft support on-boarding – that is enrolment to subscriptions offered by these companies for their mailing solutions – to many thousands of users per week globally. As such, it is obviously an advantage to choose any of these or similar other cloud computing partners with a reliable and proven track record to help create new users for organizations quickly and with the minimum of steps required in the process.

Allocation of IT Personnel

Today, technical personnel with enough expertise and experience remains valuable resource regardless of the preponderance of engineering colleges and institutions that train individuals to succeed as IT professionals. To tie down such valuable resource with relatively mundane activities such as email support rather than encouraging them to explore business critical activities, can put a strain on cash resources of a construction company. This becomes more acute in the light of the fact that the core focus of business activities of construction companies has nothing to do directly with managing IT resources on a full-time basis. As such, and in this context, it makes business sense for organizations in the construction sector to move onto cloud offerings for their mailing solutions. Technical personnel of IT departments can then be freed to perform more productive tasks befitting their skills, knowledge and interests.

Operating Latest Software

With on-premise deployment, deploying latest software and associated configurations can be an increasingly challenging exercise. While the technical complexities in such cases are sure to attract interest and attention of many a devoted IT manager, for organizations this means downtimes and loss of manhours. With cloud computing, these problems are easily resolved. Further, cloud-based software products such as Symantec MessageLabs and Proofpoint can steadily keep message filtering software and configurations updated with the latest in security patches and, thereby, deny possible opportunities for virus and malware to pass inside or infiltrate undetected.

Managing Costs Associated With Corporate Mailing Systems

At this juncture, it may not be out of place to discuss a few points that are relevant regardless of on-premise or cloud deployment of corporate mailing systems. These are as follows:

Consolidation of Mailboxes Using a Single Vendor and Single Version

The justification for having a single product version from any single vendor for corporate mailing solution cannot be over-emphasized. The fact that updates and upgrades would go ahead considerably faster given a single vendor and single product version, is enough to give this point a careful consideration in the light of operational interests and for saving productive business hours.

Identify Employees With Larger Mailbox Needs

Research has shown that not all users require identical mailbox sizes or even similar features. Based on roles and responsibilities of users, mailbox sizes and features can vary. Therefore, a carefully considered provisioning plan to allocate user-based mailboxes with relevant features can prove exceptionally useful and cost-effective.

Mail Archives to Meet Regulatory Requirements

Corporate mail archival solutions can be expensive. Therefore, identifying users whose mailboxes need to be archived to meet mandatory regulatory requirements, can be a productive exercise. In case of both on-premise and cloud computing deployments, mail archive sizes can grow quickly and to large proportions that can easily outstrip mailbox sizes allotted or initially planned.

The Case for Cloud-based Corporate Mailing Solution

The table comparing features available with on-premise vis-à-vis that of cloud computing solution speak loud and clear as to the advantages that the latter offers to organizations seeking to migrate or setup their mailing systems on cloud. However, broadly there are four different service architectures that construction companies can follow to address their respective business requirements. Certainly, a one-size-fits-all approach does not work well in all cases regardless of cloud computing's ubiquity and general tendency to save costs either early on or in the long run.

On-premise Email

Regarded as traditional and conservative, many organizations maintain that this form of mailing solution best meets their requirements on the grounds of security and data access.

Cloud Mailing Solution

For those organizations who are eager to lower capital expenditure and reduce operational costs, cloud mailing solutions can offer the best of both worlds from the point of view of email security and fees to be paid to cloud service providers per user.

Hybrid Hosted Support Services

Hybrid Hosted Support Services are a combination of on-premise and cloud computing deployment. In such cases, mailboxes are kept on-premise, while services such as archival and email filtering are hosted with a cloud service provider. Examples of cloud service providers who extend such form of services include Azaleos, Yahoo Zimbra, Iron Mountain, Google Postini and Symantec MessageLabs.

Split-domain Hybrid Mailing Approach

As the name suggests, mailboxes of some users are kept on-premise while the rest are kept on cloud. Every corporate mail user has a common email domain regardless of whether their mailboxes have been kept on cloud or on-premise. Cloud service providers often support such an architecture as this allows organizations to balance their own data security requirements with that of cost considerations.

CLOUD ERP

Much emphasis has been laid on Enterprise Resource Planning (ERP) and solutions that allow organizations to use ERP in order to manage almost all aspects of business operations. For organizations in the construction sector, and especially in India, successful ERP implementation can often translate to cost savings if not outright spark in profitability. Certainly, operational efficiency and accountability is significantly influenced through introduction of ERP. As with many other traditional IT systems that solely relied on on-premise deployments to slowly being accepted on cloud deployment platforms, ERP systems on the cloud are broadly segregated into three types.

- **Public Cloud ERP** - This form of ERP solution is by far the most mobile and flexible. It also allows for quicker implementation and extremely cost-effective. Available on subscription basis, public cloud ERP models are popular with enterprises seeking specific features from an ERP solution at reasonable rates.
- **Private Cloud ERP** - This is mostly regarded as traditional or on-premise form of ERP deployment. Large companies had often relied on this form of deployment where private data centers are setup to support ERP operations. The single biggest drawback of this form of ERP deployment is that it requires extensive manhours and large sums of money to pay for upfront and maintenance costs of sophisticated hardware, software licensing, associated supporting infrastructural charges.

- **Hybrid Cloud ERP** - This form of deployment is chosen in select few cases where an organization may be looking forward to transitioning into cloud-based deployment in a phased manner. In such situations, legacy or on-premise ERP systems may have fallen short of requirements or has grown too costly to maintain leading to an organization to explore cloud-based ERP solution.

As in the case of corporate mailing solution, the argument for transporting ERP systems to cloud holds true considering reduction of capital and operating expenditures, periodical upgrades and updates to hardware and software systems. Today, cloud computing services offered by vendors selling cloud ERP solutions feature vastly sophisticated security measures and layers to protect data and application at every stage. The only subjective issue that can override a decision to consider immediate migration to cloud-based

Table 2. Comparison of features of cloud BIM applications

Applications / Features	Autodesk BIM 360	Cadd Force	BIM9	BIMServer	BIMx	Onuma System
Clash Detection and Coordination	√			√		
Design tools	√	√	√			√
Field management tools	√					
Energy Analysis	√					√
Structural Analysis	√					
3D Visualization	√			√	√	√
Online File Storage	√	√	√			√
Online drawing viewing	√	√	√		√	√
Online drawing editing	√	√	√			√
Encrypted data	√					√
Online sharing of files	√	√	√	√	√	√
Simulations	√					√
Product Lifecycle Management	√					√
Hardware support		√	√			√
Virtualized desktop		√	√			√
Web-browser access	√			√		√
Offline access	√				√	
Mobile apps	√		√		√	√
Additional supporting apps	√	√	√			√
Private cloud			√	√		√
File management system	√			√	√	√
IFC file recognition				√		√
Change detection				√		
Free trial	√	√	√			
Free open source software				√		

ERP solution from a traditional on-premise system is that which concerns service level agreements describing data privacy and ownership of data. Considering that cloud service providers operate multiple data centers at diverse geographical locations, such data centers automatically fall within legal purview and framework of the country in which a data center may have been set up and is operated.

Advantages of Cloud-based ERP for Construction Sector

- Inexpensive - Cloud Service Providers are responsible for setting up of underlying IT hardware, software and data center premises. Companies seeking to utilize cloud-based ERP need only pay per use. No additional costs need to be paid for software licenses or any other license as such.
- Access Control - Users are allowed access to the extent requisite permissions and privileges granted by administrators from organization itself. Access to cloud-based ERP systems can take place over the Internet and on any compatible medium such as laptops, desktops, tablets or even mobile phones.
- Integration - Cloud-based ERP systems can easily be integrated with systems such as biometric devices and even CCTV monitoring solutions.
- Comparison based on Costs - Cloud-based ERP solutions are significantly cheaper in comparison to on-premise solution. The comparison table illustrating the differences between these two deployments holds true even in the case of cloud-based ERP solution as it is for cloud-based corporate mailing system.

Issues Involving Cloud-based ERP

There are several aspects and points to observe before drawing up plans to implement a cloud-based ERP solution. These are as follows:

- Identification of architecture and software licensing models.
- Annual maintenance contracts.
- Data integrity and system monitoring.
- Customization.
- Planning for future upgrades and integration with trending technologies, viz., IoT, artificial intelligence, machine learning, blockchain, etc.

CLOUD BIM: AN OVERVIEW

Although Building Information Modelling (BIM) has been trending in the construction industry in recent years, the fundamental concepts of this has been in existence for over several decades. In simple terms, BIM is an integrated information model that contains drawings, specifications of materials, details of constructions, and other paraphernalia housed in a centralized or distributed software database systems. The arrival of BIM in mainstream construction engineering has initiated a revision in design procedures and documentation. Building descriptions typically include building systems involving electrical, HVAC, mechanical and plumbing components. BIM is reliant on Information Technology for management of data and allied processes. BIM is founded on the premise of virtual project formulation first so that project planning and management framework can be put in place well before actual construction commences. BIM allows virtual integration and design in 3D. While this is certainly a definite advantage, it is not the only one by far that BIM offers. BIM enables construction teams to collaborate and analyse designs very early

Table 3. Applications with number column references

No	Application	No	Application
1	Autodesk Buzzsaw	17	Threshold CM
2	Autodesk Constructware	18	PMWeb
3	Conject PM	19	Builder Trend
4	Build Topia	20	Co-construct
5	Jonas Premier	21	Paskr
6	Mclaren Fusion Live	22	Owner Insite
7	eBID Systems	23	Spitfire Project Managemet
8	Eadoc	24	Builder Tools
9	Procore	25	Viewpoint
10	e-Builder	26	Jobnimbus
11	Built it Live	27	Accubuild
12	Aconex	28	Simplex VPO
13	Projectmates	29	Trimble Construct Job
14	eSUB	30	Prolog
15	iSqFt	31	Proliance
16	Jetstream		

Table 4a. Comparison of features of Cloud Project Management products

Applications / Features	1	2	3	4	5	6	7	8	9	10	11	12	13	14	15	16	17	18	19	20	21	22	23	24	25	26	27	28	29	30	31
Web Browser Access	√	√	√	√	√	√	√	√	√	√	√	√	√	√	√	√	√	√	√	√	√	√	√	√	√	√	√	√	√	√	√
Mobile access	√		√	√		√			√	√		√	√	√			√		√	√	√	√	√	√	√		√	√		√	√
Integration with Design Software	√																														
Drawing Viewer & Markup	√		√									√											√								
BIM Model integration	√											√						√												√	
Document Management	√	√	√	√	√	√	√	√	√	√	√	√	√	√	√	√		√	√	√	√	√	√	√	√		√	√	√	√	√
Document Editing					√																		√							√	√
Scheduling				√		√			√	√	√		√			√		√	√	√		√		√			√			√	√
Reporting	√				√	√		√	√	√	√	√	√	√		√		√		√	√	√		√	√		√	√		√	√
Drawing Management	√		√					√																							
Tender Management	√		√	√		√	√		√	√		√	√		√									√	√						√
Change Management			√																√	√		√	√	√				√	√		

on in the construction phases, and continually help designers, engineers, and architects in managing construction processes.

Under field conditions, information on project sites and prevailing conditions can be fed into BIM and detailed analysis performed on such data sets. BIM offers 4D modelling in which entire schedule of construction can be plugged in with the actual model of the construction project to help in visually identifying construction activity sequences. Often referred to as 5D, cost estimation adds another important dimension that incorporates quantities and rates of materials involved in the construction project. However, this is not all that 5D is all about. With the aid of this feature, it is possible to store such information in database systems and produce meaningful reports that can help in making pertinent financial decisions based on either perceived or computed economic factors.

Table 4b. Comparison of features of Cloud Project Management products

Applications / Features	1	2	3	4	5	6	7	8	9	10	11	12	13	14	15	16	17	18	19	20	21	22	23	24	25	26	27	28	29	30	31
Lead Management				√												√															
Contract Management				√					√																						√
Subcontract Management					√		√									√					√		√		√						
Finance Management				√	√			√					√					√			√	√	√		√		√		√	√	√
Workflow Management						√				√		√	√					√					√		√						√
Budgeting					√					√						√		√		√		√	√	√	√	√	√		√	√	√
Inventory Management					√																										
Procurement				√	√																				√					√	
Task Management												√								√						√					
Service & Warranty Management				√															√	√				√	√						
HR Management					√				√					√				√													
Material Approval			√																√	√				√							
Encrypted Data			√			√																							√		

Applications of Cloud BIM

- **Visual depiction:** BIM facilitates in intuitive visualization of even the most complex 3D designs. Projectors are often used to project 3D digital images to generate an immersive and engaging experience.
- **Synchronization:** With the aid of BIM, architects, planners, engineers and customers can all stay focused and in sync with each other to adhere to targeted delivery schedule of a project. Further BIM can help highlight errors or anomalies that can creep in from time to time during execution of projects.
- **Planning Schedule and Deployment of Resources:** BIM software are extremely useful and reliable in estimating conditions at actual field

conditions, calculating raw materials and resources, and keeping a close watch on resource utilization.

- **Setting up Sequences**: Time is a critical factor that has to be considered in any project. This can be integrated into BIM to visually analyse and take necessary optimization steps for preparing schedules and sequences throughout construction stages.

Table 4c. Comparison of features of Cloud Project Management products

Applications / Features	1	2	3	4	5	6	7	8	9	10	11	12	13	14	15	16	17	18	19	20	21	22	23	24	25	26	27	28	29	30	31
Activity Log	√	√				√										√			√	√											
Admin Control	√	√			√		√	√	√								√						√								
Online messaging		√					√		√			√																			
E-mail			√						√		√	√			√							√		√						√	
99% Uptime			√			√																							√		√
Website integration				√																											
Trade Network				√			√								√																
Contact List							√								√											√	√				
Calendar						√																				√		√			
3rd Party Apps						√			√	√		√				√		√					√	√	√					√	√
Archiving											√	√	√							√		√			√			√			
Photo tagging to Drawings																	√														
Desktop Virtualization																									√					√	
Offline Access																									√					√	
Software Installation																									√					√	
Customizable													√																	√	

- **Computing Costs:** Quantities and rates can be stored in databases used by BIM. In turn, BIM can be used to ascertain costs and effect design changes to optimize expenses both in the short term as well as in the long run.
- **Incorporating Subcontractors and Suppliers:** BIM allows to directly pull in specifications of raw materials provided by subcontractors

and suppliers. This reduces time involved, and greatly improves productivity.

- **Systematic Approach:** Using BIM it is possible to integrate all raw materials and resources under one umbrella. Equipment and their respective designs can be checked thoroughly, and conflicting areas can be resolved and addressed quickly.

Table 2 illustrates key features and differences that exist among various software offered by prominent software publishers in this field.

CLOUD PROJECT MANAGEMENT

As in the case of cloud computing deployments discussed in previous sections of this chapter, project management applications can benefit if deployed on cloud computing platforms. While applications such as Microsoft Project and Primavera are immensely useful for the tasks for which these have been designed and developed, data silos are automatically created with each user working independently of the other without being able to collaborate in real-time. With proliferation of the Internet and cloud computing, project management software have now found a new home on the cloud. The advantage of this is immediately apparent as in many cases the availability of a compatible and appropriate web application is enough for users to simply login with a selected username and associated password. Such applications can even be accessed on tablets and mobile phones either using a custom app or web browser installed on device. Some of the well-known products in this sector are Autodesk Buzzsaw, Project Mates, etc.

Cloud project management may also comprise some or all of the following features:

- Document Management
- Reporting
- Tender Management
- Finance Management
- Change Management
- Lead Management
- Contract Management
- Sub-Contract Management
- Workflow Management

- Task Management
- Procurement & Inventory Management

To discuss the features in details is beyond the immediate scope of this book. Most are, however, self-explanatory. For readers trying to grasp cloud project management, we have attempted to provide a gist of what is possible and is offered in the market. To this end, tables listing the features and comparison of products are presented in Table 3. The first table has been prepared to serve as reference to the column headings presented in the main table sections. In view of the fact that the names would be too large and unwieldy to fit in columns, the first table acts as a legend associating the names of the products with column headers and numbers.

A comparison of features of Cloud Project Management is presented in table format as shown in Tables 4a-4c.

REFERENCES

MalizaSalleh, S. Yen Teoh, S., & Chan, C. (2012). Cloud Enterprise Systems: A Review Of Literature And Its Adoption. *PACIS 2012 Proceedings*. https://aisel.aisnet.org/pacis2012/76

Purohit, G. N., Jaiswal, M. P., & Surabhi Pandey, M. (2012). *Challenges Involved in Implementation of ERP on Demand Solution: Cloud Computing. International Journal of Computer Science Issues.*

Section 4

The Future Impact of Information Technology in the Construction Sector

Chapter 8
Automating Information Flow in Civil Construction Through Internet of Things

ABSTRACT

The concept of internet of things, or IoT as it is popularly referred to, came into being when it was discovered that it is possible for computers to communicate with devices ranging from simple home appliances to industrial machinery and equipment without the need for any human intervention. The fact that IoT makes it possible for computers to simplify manufacturing processes and bring about a radical change in our daily lives, making this technology an important step forward and in tandem with cloud computing. In this chapter, the authors discuss the contributions that IoT can make in construction industry. With IoT it is possible for a machine to let a remote computer know from any part of the world its current status and physical disposition. For construction firms that need to operate machinery and equipment in distant locations, IoT makes it possible to collect monitor and manage in real-time.

It is said that a classic case of Internet-of-Things in action was that of a vending machine that was able to report the status of its stock of beverages to a computer system via the Internet. This case, which can be found on Wikipedia, is often cited as an important example of how IoT can simplify automation and take it a step forward. Today, the concept of IoT is fast gaining

DOI: 10.4018/978-1-7998-5291-9.ch008

widespread popularity for reasons related to simplification of computing processes, economies of scale, and automation.

The construction sector is one of several in which production takes place mostly in open spaces and in far-flung rural areas. Owing to remoteness of construction projects, construction companies tend to deploy a large number of personnel for any kind of construction activities be it residential or public works. A significant amount of investment is made in construction machinery, tools and equipment, cranes and rolling stock. The latter is usually used as a generic term to describe large vehicles that are either moved on rails or on road or employed for carrying large equipment and heavy industrial goods. All these, including rolling stock, are the responsibility of staff and personnel of the construction company responsible for execution of project. However, it is often found that these are subject to theft and mishandling, thus, resulting in rapid decline from the original purchase value of such equipment, vehicles, and tools. The prevalent competitive scenario in markets, both domestic and foreign, has made organizations highly sensitive to the question of cost. Construction companies, especially, have begun to sit up and take notice that allowing theft and pilferage to continue will eventually put a lot of pressure on profitable operations leading to a situation where cost overruns can make a company to close shop and declare losses.

Construction companies are also often not regarded as exemplars when it comes to balancing safety and handling of latest technology. Although, widely considered to be an indicator of economic development, the construction sector as a whole suffers from lack of optimization and personnel safety. It is this void that IoT (Li & Yu, 2011) can help fulfil an extremely important role. The construction sector is seen as one of rapidly burgeoning sectors among manufacturing industries in India. This book introduces IoT and tries to explain how IoT can help in achieving improvement in production and manufacturing processes in construction activities at remote locations. It also identifies and examines closely the set of drivers which influence approach to implementation of IoT in the construction sector. As a simple example to help follow how IoT can have a meaningful impact in day to day operations of a construction company, information and data on modelling related to building (Choung, Zihao, Yifeng, 2011) that is being constructed, energy consumption pattern, movement of machinery and personnel, and temperature fluctuations throughout a building site can be continuously monitored and kept updated for instantaneous retrieval and onward processing by the management and authorized personnel.

Construction projects today have grown more complex. With advancements in technology and mankind's insatiable thirst for improving lifestyle that was once considered beyond the reach of contemporary engineering techniques, construction projects typify the nature of safety and security concerns while maintaining targeted levels of productivity. Overcoming problems requires real-time raw data to be streamed back to decision makers. IoT makes it possible for data interchange and finer degree of control with remote appliances and machinery communicating with computers located at locations many hundreds of miles away evidencing that distances in between matter less as network capabilities and the reach of Internet grows.

The advantage that IoT is based on the feature that connected devices and machinery can not only intercommunicate but can also perform tasks based on machine learning and artificial intelligence. At the basic level, for a device or an equipment to be considered IoT-enabled, it needs to have sensors, network connectivity, and processors. Such features enable devices to sense the environment around them and either make decisions on their own or transmit information to central computing systems for instructions and relevant data to respond according to changing situations (Clark, 2016). IoT systems make for an excellent investment in conditions with fluctuating operating parameters and where decisions, if made in real-time and within a short duration of events occurring, can translate to even the most marginal increase in profitability for a construction company. Today, the evolution of IoT has brought this technology at a watershed moment where IoT-enabled devices are the norm and no longer an exception for industries that have projects in remote and outdoor locations.

IoT brings to the table several advantages from which construction companies can move forward in order to control costs and boost productivity. Fuel-monitoring systems and vehicular operational status monitoring solutions are standout examples of IoT in action in remote project locations. Already several construction giants, for example Simplex Infrastructures Limited, use IoT enabled fuel-monitoring systems that provide near real-time feedback on fuel consumption data. As its name suggests, IoT relies on the Internet for communication. Communication being a vital ingredient in the overall scheme of a typical IoT solution, there are applications where Internet can be temporarily bypassed in favour of local area networks (LAN) and wireless area networks (WAN). Such forms of networking can prove to be useful for a project site can have its own central servers communicating with IoT-enabled devices and equipment over either LAN or WAN, and then process data and information for onward transmission over the Internet to primary servers located at city offices of a construction company. Be as it may, and whatever form IoT may take at field level, it goes without saying that IoT has transformed the manner in which construction companies can exert greater control at a micro level and at the actual production stage. As such, IoT thus helps organizations to automate tasks that would have other required costly manhours for gathering data and initiating action that, in turn, can cascade into further series of data collection and responses.

IOT

Internet-of-Things is described as an environment that involves an object that has sensors, storage and microprocessors for collecting, assimilating, transferring, and taking action based on data collected from immediate surroundings in which an object is placed. For such a device to be considered part of IoT system, it should be able to transmit and receive data wirelessly over designated networks and/or the Internet. An IoT-enabled object need not be a computer in the physical sense as we are accustomed. It may be any form of regular commodity or general appliance, e.g., TV sets, washing machines, indoor lighting system, faucets and washroom fittings, vehicles, machinery, logistical equipment, etc. One of the more advanced examples of IoT in action is how Amazon, one of world's larger retail giants, handle packages and consignments and display tracking status in almost real-time to its customers.

HOW IOT ENABLED CONSTRUCTION COMPANY WORKS?

A hypothetical example of IoT system in action after a construction company decides to upgrade its systems and implement it, can be described as follows:

The example involves a construction supervisor who begins a typical day at the beginning of a week. He boards a company-provided transport that automatically logs in and tracks vehicular movement reporting the data in real-time to central servers and connected software time clocks. This automatically enables the company's systems to record consumption of fuel, stops made, if any, and general speed and direction of travel. More sophisticated systems can make use of Google Maps to plot movements collected from data received and plan optimized routes considering daily commute routine of the supervisor. Once the vehicle carrying the supervisor arrives at site, geo-fencing systems and on-board IoT devices inside the vehicle communicate to log arrival. Simultaneously and subsequently, information and supervisor's availability and presence in site is transmitted to authorized personnel.

A supervisor can have access to locational information of heavy equipment and personnel at site. Monitoring of data and locational awareness of assets and human resources can go a long way in reducing losses. Routine tasks that can be performed through automation can be controlled from central servers located at a Data Center several hundreds of kilometres away. Operational processes, either segmented or integrated, such as orders, deliveries, inspections, quality controls and invoicing can be optimized to adhere to budget and costs. As such, streamlining daily operational procedures is possible and facilitates in systematic record-keeping and coordinating administrative activities efficiently. IoT enabled systems can help predict potential issues before these can snowball into major problems. This feature is important as costs can be kept to a minimum and sudden escalations can be avoided. Analysis of operational downtimes and data related to equipment use helps supervisors and managers to coordinate maintenance schedules and predict requirement of spares. This, in turn, helps in maintaining a low inventory at a time when inventory carrying costs can pose to be a significant issue in keeping operational costs as low as possible.

While the above example is a snapshot on the capabilities of IoT and what is possible using this technology, it is important for us to discuss and make a note of the areas in which the emergence of IoT can have a lasting impact:

Equipment Repairs

Today, many heavy construction equipment and vehicles are being equipped with sensors that broadcast data on operating parameters. The broadcasts can be picked up wireless sensors that act as repeaters to boost signal transmission and forward the messages received to a central location within a site for onward processing by central servers at data centers. In certain cases, on-board sensors can be coupled with processors, GPS units, and mobile technology to transmit and receive messages directly. This capability, although, more expensive is fast growing as the preferred choice for managers and project planners seeking to leverage this technology to dynamically tune operational activities remotely. Certainly, this poses significant advantages in that manpower and resources can be deployed more accurately and in a timely fashion to either coincide with an issue that might be developing, or before an issue can develop. Such measures can greatly enhance operational and situational awareness leading to prevention of costly delays.

Preventive equipment maintenance is often overlooked by managers and supervisors at construction site. This approach stems not from their ignorance or indifference, but rather as a result of mounting pressure of immediate and vital activities that require resources and manpower to be deployed. Sadly, lack of preventive maintenance can lead to some of those very issues that preclude supervisors and managers from initiating preventive equipment maintenance in the first. This is a cycle that construction companies need to break out of in order to minimize rate of breakdowns and control resultant loss in revenue. With the help of IoT, it is possible for machinery and heavy equipment to report diagnostic information on their own and automatically. With the help of this automated delivery and system-controlled responses, routine as well as preventive maintenance activities can be scheduled in far less time and involving optimized amount of resources.

An IoT based system that monitors and manages heavy equipment and vehicles at constructions sites can offer a varied amount of information at a single dashboard that can be viewed by authorized personnel with access to the Internet and using laptops, desktops, or even mobile platforms. Information such as make, model, serial number, warranty information, manuals, records of maintenance can all be accessed using a portal available 24x7.

Often IoT based systems that cater to management and monitoring of construction machinery and heavy equipment, are paired with a scheduling software. The purpose of scheduling software, as its name suggest, helps plan preventive maintenance. Using scheduling software, several activities can be scheduled such as preparation of orders, setting frequency of repairs, creation of checklists, incorporating images and documents, etc. Such scheduling software sends alerts to designated personnel informing them of activities and action to be undertaken. Activity options can be as follows:

- Scheduling by dates.
- Scheduling by meter readings or by equipment usage.
- Scheduling by approximate usage monitored by meters.
- Conditional-based scheduling.

Tools and Tracking

Aside from heavy machinery and vehicles, tools and smaller equipment can also be made part of an IoT network. This reliance on IoT can not only reduce the number of instances when such are misplaced or lost, but can also minimize costs on account of replacements and loss of productive hours. Considering that a construction project may involve many hundreds of assorted tools and smaller equipment, chances of loss or misplacement are high. Implement an IoT based system to track equipment location can prove to be of value.

Building Information Modelling

Building Information Modelling accommodates IoT and embedded sensor technologies. Building Information Modelling is relied on extensively by architects, construction engineers, and project planners to design smart buildings. A generic concept of a "smart building" is one that has sensors deployed and embedded throughout its structure. These sensors stream back vital data and information to centrally located computers which can then analyse the data to facilitate and report on suggested course of action to be initiated from time to time. The data that such sensors collect can include several parameters including immediate environmental information on weather, earthquakes, etc. The ability to stream data and information by the sensors assist designers and engineers to create better designs that can withstand vagaries of nature and last longer.

3D LASER SCANNING

Building Information Modelling also includes 3D laser scanning technology. This laser technology can help optimize labour expenses, help in estimating cost and resources expenses, as well as reducing the overall time required. Laser scanning is relied on by civil engineers and contractors to make accurate calculations in projects that can also include those that involve restoration and retrofitting.

Laser scanning technology is currently the most accurate method known for measuring distances. In a technique known as LIDAR (Light Detection And Ranging) which involves emitting laser at points very rapidly to calculate the accurate distance. Construction companies use tripod mounted LIDAR instruments to measure distances while conducting ground surveys. LIDAR instruments are very accurate and can measure distances ranging from 100 to 200 meters. What makes LIDAR instruments indispensable with civil engineers and surveyors is the high rate of accuracy and extremely rapid scanning rates. It has been observed that there is no marginal error of not more than two millimetres, plus or minus, from actual physical distance for every 100 meters being the distance from the instrument itself to object.

Construction companies favour LIDAR instruments over traditional survey methods that rely on physical measurements and mapping. While these methods are reliable, these are not as accurate as LIDAR and involves a lot more effort and human resources to conduct accurate surveys approaching close to the level of LIDAR. With evolving drone technology, where small remote-controlled flying crafts can cover large and rugged terrain staying aloft for several hours at a stretch, cameras and 3D laser technology has found a new platform for use. The use of 3D laser scanning enables engineers to create detailed surface models of almost any object or terrain. Mounted on drones, cameras and 3D laser instruments help construction companies to come up with detailed maps with accurate contour surveys covering vast areas in shortest possible time.

Laser scanning technology has been found to be very useful where renovation and restoration of heritage buildings are concerned. Architects and civil engineers rely on laser scanning technology to determine the locations and plan installation of fire protection and HVAC systems. The advantage that laser scanning is that engineers can plan renovations and restorations without causing damage or any structural issues to existing structure of building. Software developers can program software code to allow interaction with laser scanning and LIDAR instruments. Major software vendors now offer software and software addons that allow architects and civil engineers to plan and design construction projects with LIDAR instruments seamlessly integrated with 3D CAD and modelling software. Such modern 3D CAD modelling and BIM software allow the cloud of point data generated by LIDAR instruments following scans, to be stored and processed as 3D images. This feature is extremely useful in cases where architects need to plan and design structural upgrades based on drawings of existing buildings and structures.

Fleet Management

Fleet Management involves tracking vehicles and collecting data on operating parameters. Vehicular tracking and monitoring is made possible through use of embedded sensors and processing chips that transmit information related to engine operations, R (RPM), current torque generated, fuel consumption rates, temperature and pressure, current speed, run times, idle hours, and location of vehicle. Fleet Management software typically provide analytical dashboards and assorted reports to assist managers and supervisors to have a finer degree of control over vehicles that are in use at sites. One of the more significant advantages of Fleet Management software is provide data and analytics in near real-time that translates into more effective preventive maintenance schedules. Such preventive maintenance schedules when predicted by Fleet Management software translates to optimize costs related to spares and human resources required for conducting repairs and maintenance to costly equipment and machinery.

Fleet Management software is one of stellar examples of IoT technology in practice commercially as well as in industrial sectors. Since, managing even a small fleet of vehicles at construction sites can be a problem, such software systems with IoT technology can help automate operational activities and detect problems well in advance. This, in turn, can improve decision-making process affecting projects and operational activities of construction companies. In addition to IoT, Fleet Management software depend on telematics and global positioning systems. However, at remote project sites where telematics can be affected by a combination of rugged terrain or other technical factors, the usefulness of fleet management software can be degraded. In such situations, most modern fleet management systems offer a feature that allows recording and storing of data onboard even when signal transmissions are not possible. Once a vehicle is within range of mobile networks or wireless area networks allowing it to receive and transmit data, fleet management system automatically moves into gear to push data to cloud or to central servers installed at data center used for hosting fleet management system software.

Many proponents and supporters of software such as Fleet Management are of the view that similar to buying insurance cover for vehicles, installation and commissioning of Fleet Management system is most essential. Under actual conditions, the advantages and benefits of operating a Fleet Management software have been observed across the globe where vehicles are operated in remote and far-flung areas well away from central offices or centralized transport control locations.

Following are the advantages of Fleet Management software:

- Tracking location of vehicle at any given point of time.
- Reducing fuel wastage.
- Increase chances of Recovery of Stolen Vehicles.
- Resolve billing disputes.

Wearables

Wearables, as the term implies, is a form of technology that includes electronic devices that can come in the form of sensors and transmitters embedded in clothing, or as micro-computers complete with sensors and GPS features that can be worn on one's person. Such devices can come in a variety of shapes and features ranging from smart watches, AR goggles, smart helmets, smart wristbands, etc. These devices can gather and transmit data collected off of an individual or one's immediate surroundings. One of the first successful wearables was Google Glass. Augmented Reality (AR) was first popularized by Google Glass although the technology itself was not unknown before that was introduced. What makes wearables important is the fact that it allows for processing information and data at a more personal level.

Wearables have probably more recognized in the form of Smart Watches that can track human pulse rates, location, exercise metrics, etc. These features have popularized wearables as devices that can help detect initial problems with health, and also provide a form of detailed analytics for the casual observer. In construction industry, wearables are extremely useful in determining personnel position and collect data in their immediate vicinity. In sites and remote locations where construction is in progress, managers and supervisors need to know situations and issues as and when these are noticed. Wearables are an effective means of collecting data on potential issues and safety hazards. Data collected through networked wearables can be transmitted almost instantaneously to connected users and central systems for immediate attention of all personnel affected either directly or indirectly.

Smart Watches offer features and functionalities that extend more than time-keeping duties. These function as miniature computers strapped to one's wrist able to communicate with either a Bluetooth-paired phone or act independently. Smart Watches can accept voice calls, transmit and receive text messages, and communicate independently over mobile networks. What makes Smart Watches extremely useful in remote areas are their versatile features combining that of functions of a typical phone with geolocational information and other relevant personal data such as health, steps taken, etc.

Wearables can also be in the form of Smart Clothes. Although still in nascent stage, this form of wearables is not obvious when compared to Smart Watches. Although the concept has been in existence for some time, the use of this technology is yet to attain a level of commercial success as that of Smart Watches. Similar to Smart Clothes, researchers are also conducting studies into advancing IoT technology to shoes. In future, Smart Shoes may be able to effectively gather and collect data related to field conditions that may not be otherwise be possible through other forms of IoT technology. Smart Glasses are fast gaining as the next generation of IoT devices. The basic concept of Smart Glasses stems from heads-up-display units (HUD) that military pilots are equipped with. Smart Glasses are capable of collecting visual data as well as other information through tiny sensors embedded. Smart Glasses support wireless network connectivity making these extremely useful to transmit visual images in real-time to central systems. Users of Smart Glasses can view information right in front of their eyes on their geographic location, equipment information and data that can be streamed back to them from central servers, and other important data required for performing important functions.

Architects, engineers, supervisors, foremen, skilled workers and even unskilled labour resources can benefit from Smart Glasses. Personnel supervising construction activities, and architects, can view in real-time the progress of different activities at remote project sites. Immediate remedial action can be taken, if necessary, or additional information can be passed on to workers engaged in construction activities. Project Managers, contractors, engineers, health and safety supervisors can also determine the state of health of workers through instantaneous analyses of eye movements that can indicate tiredness, deterioration of health, or in case where any worker may be deliberately avoiding set tasks or activities. Smart Glasses are extremely useful in cases where ensuring safety of personnel is a concern in and around hazardous construction zones such as drainage works, pipeline projects, bridges, tall towers, etc.

Augmented Reality (AR)

The concept of Augmented Reality (AR) stems from the fact that computer technology is able to reproduce images and figures produced from data received from various sensors. These sensors may be in the form of optical, audio, temperature and many more. In AR systems, computer generated visual data can be displayed straight on to special liquid crystal displays (LCD) or heads-up display units (HUD). The primary purpose of AR systems is to provide users with additional and more functional details that may not be obvious or evident to users in actual conditions. For example, a construction worker with an AR headset can obtain information on dimensions of piling streamed to him while he is actually carrying out the work. A similar example could involve an engineer supervising construction of a stretch of road with information on temperature, humidity and other relevant information displayed to his AR headset.

Supply Replenishment

The process of supply replenishment forms an integral part of an Inventory Management system. IoT technology can be integrated in the form of consumables and spares affixed with radio frequency identification (RFID) tags. A RFID scanner can be used to scan the rates of movement and consumption and transmit data to central servers. An inventory management system can then initiate supply for more items depending on quantity available and the rate of consumption of items. As such, overhead costs for keeping a constant level of items at stock can be reduced and inventory can be maintained on the basis of consumption rates and availability of items.

Intelligent Prefabrication

The use of prefabricated building parts and items have long since proved that this can save costs over traditional building methods. Further, the use of prefabricated building components greatly lower amount of waste produced at construction sites. This approach requires certain complex manoeuvring involving calculations related to quantities required and physical availability at any given point of time. In such cases, too, IoT enabled systems involving RFID can simplify logistical movement of prefabricated items. This enables managers and engineers to plan construction and take into account any delays either during transportation of prefabricated components on way to construction location or during actual construction process.

Green Building and an Approach to Efficient Energy Management

Smart Buildings are fast gaining popularity for various reasons. One of the primary factors beings power consumption. Building infrastructure involves consumption of electrical energy to drive pumps, lifts, escalators, lighting, HVAC, surveillance systems, etc. Embedded IoT devices can help provide continuous or steady stream of information on occupancy, load factors, and environmental data to central servers from where a software system can control flow of electrical power to areas where required and turn off supply when not needed.

Construction Management

Construction Management now holds enormous significance for construction companies. Time being the essence of any successful project, an improper construction management can cause serious damage to profitability and cast shadow over survivability of a construction enterprise. IoT has an extremely important role to play in construction management. Sensors and micro-computing devices coupled with antennae can transmit vital data and information to central servers continuously. As a result, construction companies can monitor and plan consumption of energy and inventory more accurately.

TRACKING HUMAN RESOURCE HOURS

Tracking of human resources and hours spent in activities related to a construction project can be a major challenge. Labour contractors employed are often at odds with construction companies employing their services when it comes to human resource hours billed. Smart Wearables and IoT can greatly simplify tracking of personnel and, thus, lead to transparent billing.

Structural Health Monitoring (SHM)

Structural Health Monitoring (SHM) is a fairly new form of monitoring approach to sensing structural condition of buildings. This approach offers a more detailed overview than visual inspection for detection and maintenance of civil structures. SHM enables maintenance personnel to have critical insight on structural loads that buildings, bridges, towers, etc. are subjected to. Wind, weather factors, earthquakes, regular stress and strain lead to tremendous pressures applied on buildings, bridges, towers, etc. With the help of sophisticated measuring systems such as GPS, terrestrial laser scanning, vision-based systems, Doppler meters, strain sensors, displacement transducers, accelerometers, etc. SHM systems are able to provide detailed reports on structural condition of any large civil construction well in advance. SHM systems also use fibre optic sensors to measure strain and forces exerted on columns in high-rise buildings.

SHM involves monitoring various aspects that can affect any structure. These are as follows:

- **Mechanical:** cracks, deformation, displacement, loads, strain and stress
- **Physical:** humidity, pore pressure and temperature.
- **Chemical:** chloride penetration, rebar oxidation, steel oxidation, sulphate penetration

The primary purpose of Damage Identification is at four levels. These are as follows:

Level 1: Determination that damage exists
Level 2: Determination of Level 1 added to determination of geometric location of damage
Level 3: Assessing level of damage severity in addition to Level 2
Level 4: Completion of all assessments at Level 3, along with the forecasting is remaining service life.

SHM systems use a variety of sensors. Broadly, these sensors can either be passive or active. Active sensors are capable of receiving as well as transmitting signals. Passive ones are only capable to receiving signals. Active sensors make use of several materials of which piezo ceramic is considered useful in the construction of such sensors.

Table 1 lists several types of sensors used and respective functions.

Table 1.

Vibration sensors	Vibration causes cracks and deformities to take place. These sensors help detect vibration.
Acceleration sensors	Useful in determining the effects felt during an earthquake, acceleration sensors play an extremely vital role.
Load Cell sensors	Load-bearing capacity is an important metric that construction engineers need to monitor. Load Cell sensors are useful in reporting sensory information related to quantum of loads that a structure is being subjected to at any given point of time.
Wind sensors	Wind forces are known to cause serious destabilizing issues with structure of building. Wind sensors are mounted, therefore, to gather wind speed and direction and transmit these data to central servers.
Pressure sensors	Similar to load cell sensors, pressure sensors are deployed to collect data on pressures that structural areas are being subject to at any point of time.
Deformation sensors	Deformation sensors combine features of displacement transducers and tilt meters. The purpose of these sensors is to report the extent to which a structure is deformed. Data from these sensors help architects and engineers make better structural designs by taking into account the forces causing deformation beyond the parameters designed for initially.
Temperature sensors	Temperature sensors report data on temperature inside a structure and that in the immediate environment or surrounding.
Oxidation-Reduction Potential (ORP) sensor	As one of the key sensors that collect data on chemical and material composition changes inside a structure, these sensors help gauge the extent to which moisture and humidity has penetrated at particular points and positions of a structure.
Corrosion sensors	Corrosion is perhaps the single-most deciding factor leading to deterioration of reinforced concrete. Architects and engineers rely on corrosion sensors to not only carry out preventive maintenance, if necessary, but also examine possible reasons for corrosion to set in at various stages of a structure's lifecycle.
Strain sensors	Strain sensors are a combination of the various pressure and load sensors discussed above. These sensors are widely used to monitor pressure on long term basis on rafts and piled rafts.

IMPACT OF IOT IN CONSTRUCTION SECTOR

There is no denying that IoT today has a vital and an integral role to play in construction sector. There are several tangible benefits, not to mention indirect and intangible benefits, to implementing IoT at project sites and far-flung remote work areas.

- **Remote Operation -** Combined with Cloud Computing, IoT provides an almost unparalleled advantage in taking immediate action on developments taking place at remote construction locations. One of the disadvantages that managers and project planners face in managing construction projects is the delay in accessing data generated at site. IoT obviates the use of human messengers or even other information technology methods in certain cases.

- **Improving Quality Monitoring Processes -** IoT and embedded devices can constantly report back the state of health of machinery, vehicles and equipment. This allows managers and supervisors to carry out far more effective preventive maintenance schedules than would have been possible. Speculation and/or routine periodic checks that can involve extensive hours and resources to be deployed can easily be avoided or minimized by implementing IoT.

It is undeniable that Information Technology is fast revolutionizing the manner in which industries and even individuals work towards enrichening the human civilization. Construction engineering is widely regarded as one of the ancient engineering disciplines to have ever gone through a constant state of change towards perfection. In this quest, humans have now found themselves another ally in the form of a revolutionary thinking in Information Technology. And that thinking or paradigm is IoT. Construction companies in future shall grow to be more reliant on IoT than ever before. While reliance on humans is not going to fade away any time soon, the emergence of IoT promises to offer profitability by optimizing time and resources for industries, engineers, architects, and end-users alike.

REFERENCES

Choung, G., Zihao, L., & Yifeng, Y. (2011). *The Research and Implement of Smart Home system based on Internet of Things*. Electronics, Communications and Control. doi:10.1109/ICECC.2011.6066672

Clark, S. (2016). *A Smart Wearable for Health-care*. Available: https://www.techbeamers.com/introduc-ing-welt-belt-a-smart-wearable-for-health-care/

Li, B., & Yu, J. (2011). Research and Application on the Smart Home based on component technologies and Internet of Things. *Procedia Engineering*, *15*, 2087–2092. doi:10.1016/j.proeng.2011.08.390

Chapter 9
Streamlining Business Operations in Civil Construction Using AI

ABSTRACT

The concept of artificial intelligence stems from the approach to empower machines to act independently of their human masters. This has been found to be extremely useful in situations where routine or repetitive actions are required to be taken based on given data patterns for which complex algorithms and decision-making processes are required to be executed. In such cases, AI has been proved to be effective in replacing humans. While instances of AI substituting humans may appear to pose potential threats to jobs, induction of AI in manufacturing processes can reduce instances of accidents owing to human fatigue or otherwise threat to life and limb working in dangerous environments. In this chapter, the authors discuss the role of artificial intelligence in building information modeling and knowledge management. They also explain how AI can be used as a building block for a safer and productive work environment.

Even a few decades ago, the concept of Artificial Intelligence (AI) captured the imagination of the public not as a technological implementation but as a fantasy to be enjoyed within the realms of cinemas and novels. After nearly two decades past the turn of the century, AI is no longer a fiction or product of fertile imagination. It is a tangible product of science, mathematics and

DOI: 10.4018/978-1-7998-5291-9.ch009

information technology that can be packaged and sold. AI, today, can be found in devices that are sold by Apple and Amazon in the form of "Siri" and "Alexa", respectively. It is also to be found in Android and Apple iPads, iMacs and iPhones. The role of AI (Russell, Norvig, 1995) in construction industry in the present context cannot be stressed enough. Aided by AI, construction companies can now plan for a better supply chain management, enhance productivity, improve quality while maintaining high safety standards at all times. Research studies conducted indicate that fuzzy logic (FL), artificial neural networks (ANNs), genetic algorithm (GA), and fuzzy genetic (FG) can yield promising results in the context of construction materials and structural mechanics Combined, these make up AI as it is known and widely used in the field of commercial, industrial and even home-use applications.

Artificial neural networks (ANN) – considered integral to AI – have been developed to a stage where volumes of data can be collected, analysed, memorized and actions can be taken independently by machines to respond to a variety of changes. ANNs can carry out extensive numerical computations based on experimental as well as data collated in real-time. Research and field trials conducted using AI and ANN have shown its usefulness in designing, assessments on structural damage, risk analyses, construction management, fleet and logistics management, and many more vital functions in the field of construction and civil engineering (Schalkoff, 1990).

WHAT IS ARTIFICIAL INTELLIGENCE?

In simple terms, AI is a computer program that can think, learn and act as naturally as a human would. While human beings have certain limitations in processing vast bulks of data and information, AI has no such problems as it uses computing resources of several dozens or even hundreds of computers interlinked and working together. Although, AI is yet to reach a point where it can completely take over the thoughts and functions of humans, the idea itself can be credited to the famous mathematician and computer programmer, Alan Turing. As Jack Copeland, Professor of Philosophy and noted historian on Alan Turing, in his book "The Essential Turing: Seminal Writings in Computing, Logic, Philosophy, Artificial Intelligence, and Artificial Life: Plus The Secrets of Enigma" states, the term "Artificial Intelligence" first featured prominently at a conference held at Dartmouth College, New Hampshire.

The title of the conference itself – *The Dartmouth Summer Research Project on Artificial Intelligence* – hosted the first ever mention of AI as we now know it today. Professor Copeland asserts, however, that the famous British mathematician used a similar but different term. The term was "machine intelligence"; not altogether misplaced since that is essentially what it is today. AI technology primarily involves silicon-based devices and inorganic substances. It, therefore, appears to be a term most suited to the current scenario and research developments that are taking place in universities, industrial and research establishments. In the UK, the term "machine intelligence" is often used interchangeably with AI.

In every way that AI is perceived, examined, reviewed and deployed, the contribution of Alan Turing is evident. It was Turing who, in 1948, described how a machine can be made to perform certain functions of the human brain. Although, his initial essay on AI in its nascent stage found no immediate followers and supporters, his next survey on "Intelligent Machinery" certainly did. Professor Copeland describes this work as the basis for future researchers and the way AI has evolved over the decades and well into 21st century.

The term "Intelligent Machinery" can be viewed as being synonymous with "Machine Learning". Although, there is a sizeable proportion of commentators who argue that AI should be viewed as a technology, there are those who suggest that an "intelligent" machine may not have human emotions or traits, e.g., rationality, ethics, morality, etc., required for classification as an example of AI. Today, AI is an assimilation of technologies and infusion of machine learning, natural language processing, artificial neural networks and inference algorithms. Computers are programmed to acquire AI in two different ways. The first being "Machine Learning". In this mode, computers are fed volumes of assorted data. Software programming is then used to train computers to take action and respond to developing situation based on conditional factors. For example, a computer can be trained to respond to multiple patterns and act based on if-then conditions triggered by complex algorithms.

The second route by which computers can acquire AI is through simulation of human brain neurons. Data and information stored over a period of time is constantly evaluated by computers and fine-tuned. Computer software can be programmed to generate improved versions of their own code. This is termed as "Deep Learning". Self-driving cars are equipped with Deep Learning to distinguish and recognize traffic pattern, road signs, and even take action based on predictions that their onboard computers come up with. Self-driving cars are no longer limited to being just research projects (Adeli, Hung, 1995). Car companies are actively pursuing commercial goals to popularize autonomous self-driving cars. One of the world's largest software manufacturers, Google, began its own self-driving car project. Today, it has morphed into Waymo – a company with the goal of popularizing autonomous self-driving cars in the US.

Figure 1. Artificial intelligence (AI), machine learning and deep learning

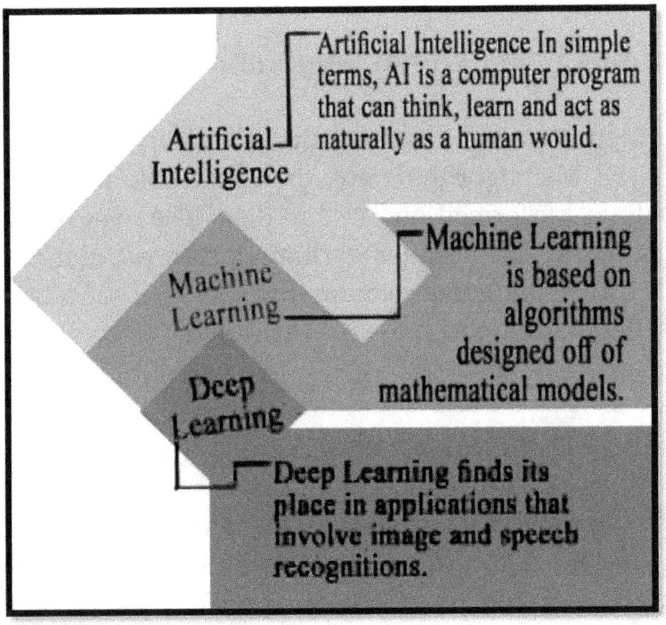

CONSTITUENT AI TECHNOLOGIES

The quest for defining and refining AI has led several interlinked technologies to be designed and developed. Some of these technologies have gained a firm foothold in their own right. While it could be argued that these technologies in isolation would not be a complete description of what AI was first envisaged, nevertheless, these are technologies that are the fruits of deliberation of many a researcher, academician and computer programmer.

Machine Learning (ML)

Machine Learning is based on algorithms designed off of mathematical models. The objective of such algorithms which are run on computers is to forecast and carry out predictions based on data sets that are similar

Machine Learning algorithms are classified as either supervised or unsupervised:

Supervised Machine Learning Algorithm

This type of algorithm uses data sets that are clearly labelled. Computers using supervised ML algorithms are, therefore, able to make predictions based on analyses performed on existing data. The data used for analyses behind prediction and forecasting algorithms are known as training data sets. Supervised ML algorithms gain accuracy progressively as more training sets are introduced. These algorithms are fine-tuned by comparing predicted output with the actual outcome.

Unsupervised Machine Learning Algorithm

By contrast, unsupervised ML algorithms are deployed on data that is neither available in labelled or classified form. Studies conducted using this type of algorithm return functions to described underlying structure that make future inference possible from data that are not labelled. Systems that employ unsupervised ML algorithms conduct autonomous improvements by exploring newer data sets.

Semi-supervised Machine Learning Algorithm

As the name suggests semi-supervised ML algorithms use data that can be either labelled or unlabelled or even a combination of both. Usually for such algorithms to be trained, a small proportion of labelled data is taken along with larger amounts of unlabelled data. Sem-supervised ML algorithm is especially suited in situations when available unlabelled data needs additional resources in order to build up training.

Reinforcement Machine Learning Algorithm

This form of ML algorithm is based on trial and error approach. Reinforcement ML algorithms interact by acting on responses and analysing feedback received. This type of algorithm functions on the basis of trial and error, and is dependent on consequent rewards or errors. In case the algorithm is able to predict correctly based on a given data, it collects a reward and, therefore, is able to learn and develop recognition patterns. Machine Learning of today is an integral part of broader AI framework. Additional cognitive technologies combined with ML algorithms are pushing the frontiers of AI. Despite the obvious drawbacks involved in extensive training and time involved initially, AI can not only help predict situations helping businesses turn profitable but also assist in creating a safe and healthy work environment.

Cognitive Computing

Cognitive Computing involves advancing Machine Learning with a number of large data sets. The aim of this approach is to simulate the way humans think and take decisions. Training of such systems usually involves reproducing human ability to discern and categorize based on conditions.

Deep Learning

Deep Learning involves using neural networks to produce analytical results that predict outcomes. It is the use of neural nets or neural networks that separates Deep Learning from Cognitive Computing today. Deep Learning finds its place in applications that involve image and speech recognitions.

Predictive Application Programming Interfaces (API)

A predictive API is dependent on AI to deliver predictions given a set of data sets. The output returned is in the form of standardized sets.

Natural Language Processing (NLP)

This technology has enabled computers to process human languages whether written or spoken, and produce speech and writing after determining the context in which communications take place.

MACHINE LEARNING IN CONSTRUCTION SECTOR

Many experts argue that Machine Learning is an extension of AI that allows computers and machinery to autonomously improve their own decision-making processes through training and learning from available data, and without any additional human intervention. In other words, the concept of ML requires that computers understand data coming through in real-time and take decisions autonomously and without humans having to program them every time a decision is needed to be made.

Machine Learning tools are able to analyse and sift through data independently. Using output tested and based on existing data, ML algorithms are progressively able to improve their ability to make better predictions. To corroborate and, thus, fine tune their powers of decision making, ML algorithms benefit from human intelligence to enhance predictions.

In the context of construction sector, ML algorithms are of value as this sector is not only labour-intensive but also involves design and planning stages which can take up a lot of time. The work involved in design and planning stages demand extensive number of months to years of manpower from teams of dedicated and qualified experts and professionals. While the work that usually gets done by such teams are vital, most of it are conducted through constant human intervention and are, thus, tended to be regarded as monotonous. Further, raw data received in the form of information may reach designers and architects in fragments with little or no continuity assurance. Errors in design and construction phases often, therefore, creep in. Given that construction companies have now to contend with diminishing profit margins in the face of stiff competition, inefficient designs and poor cost estimations through improper planning can put the very survival of a construction company at stake.

AI and Machine Learning together promises to radically alter the way construction companies approach issues related to workflows. AI and ML are capable to classifying and predicting BIM elements in workflows that can eventually lead to process optimization. There are several examples of how AI and ML can help bring about radical change in workings of a construction company. These are as follows:

1. **BIM classification:** Cost estimates for projects are worked out on the basis of bill of quantities. BIM elements reflect bills of quantities of projects. In such cases, Machine Learning helps by evaluating several

associated BIM attributes that include geometry and topology by classifying BIM elements. During this process, ML algorithm may require human intervention to classify elements for which the algorithm has no prior classification definition stored.

2. **Knowledge Engineering and Management:** Construction companies are heavily reliant on experts and professionals with specific knowledge and skills that are in demand in this sector. In order for AI to be more relevant and practicable, storing and cataloguing information and expertise of skilled human professionals are essential. Stored and catalogued data sets can then be used to train AI that can, in turn, autonomously approach workflows based on rules and predictive analyses. Knowledge repositories, if created in such way, can significantly curtail cost and many hours of intensive work involving acquiring, removing inaccuracies and labelling data in a structured form.
3. **Risk Factors:** Project Managers can design and predict a better yield using machine learning. Enormous volumes of data can be processed systematically instead of relying on guesses and estimations conducted without significant basis.

The above summarizes how AI can add competitive advantage to a construction company. With sustained and systematic use of ML, AI algorithms can only improve leading to progressively higher prediction rates. Projects costs and schedules can be streamlined through better AI predictive approach. This, in turn, affects enhanced bidding and, thus, working towards profitability and sustainability in competitive markets.

CIVIL ENGINEERING AND INTELLIGENT OPTIMIZATION METHODS

Researchers have described AI as a systematic approach to well-defined system of classifying how human intelligence works. From the days of Alan Turing through to the modern age, AI technology has been developed and adapted to be used in several applications. Industry experts and researchers agree that AI has the potential to effectively tackle many issues and problems affecting construction sector.

Evolutionary Computation

Evolutionary Computation (EC) is regarded as sub-topic of AI. Using iteration that resembles biological evolution to reach a particular biological refinement, EC has been a favoured approach in case of many applications in civil engineering that require solving of complex problems and optimize accordingly.

Genetic Algorithms

Genetic Algorithms (GA) are considered to be one of the important evolutionary algorithms. This algorithm simulates evolutionary thought and principles as set out by Charles Darwin for the natural world. GA has been accepted in civil engineering and construction sector as new mathematical models and tools are developed. GA functions on the basis of selection of the fittest approach to optimization. Similar to what Charles Darwin had posited on survival of the fittest for organisms living or extinct, GA analyses are conducted taking into account training data sets that offer better accuracy and discarding those unsuited. This process is conducted repeatedly with corresponding results evaluated. GA is a favoured approach when it comes to structural optimization and for allocation of resources to solve issues related to buildings. GA is also used where infrastructures for roads and waterways are required to be constructed efficiently. Proponents of GA argue that this type of algorithm is well suited to conducting analysis data related to construction of long suspension bridges that are required to carry enormous loads.

Artificial Immune Systems

A mix of genetic algorithm and least-squares method is used to determine structures that are feasible and to seek relevant structural constants. This relatively new approach aims to address the drawbacks of traditional as well as artificial neural networks

Swarm Intelligence

Swarm Intelligence has its roots embedded in metaheuristics. Evolution of solutions based on available data sets and then analysing through interactions within an environment forms the basis of such algorithm. Although, some researchers are of the view that it does not on occasion provide optimum solutions, nevertheless, Swarm Intelligence has shown much promise and is active in many sectors.

Particle Swarm Optimization

Based on population-centric global optimization approach, Particle Swarm Optimization involves moving individual solutions, termed as particles, through a layer of hyper-dimensional search space to identify optimum solution. Each solution, or particle, has its own position and velocity vectors that adjusted at the every iteration. The adjustments are conducted by learning process involving local best solution found by a particle and by contrasting the finding to current global best solution as determined by whole swarm to which the particle itself belongs to. This form of algorithm is considered suitable in case of complex systems and in case of solving computation problems involving social networks.

Ant Colony Optimization

Ant Colony Optimization replicates behaviour as exhibited by real ants. Ants living in colonies of their own communicate among themselves using pheromones. This helps the ants to carry out such sophisticated activity such as discerning shortest route from their nest to locations where food is available. Several researchers have examined this algorithm while researching problems related to supply chain optimization. The objective being to minimize total cost while maintaining total lead time for delivery within due dates. This is especially relevant in the context of construction industry where supply and logistics are critical elements and have a decidedly marked effect on overall profitability of construction projects.

Neural Networks

A sizeable number of civil engineering software applications that use neural networks are based on back propagation algorithm which is considered simple. This algorithm itself is based off of the development work carried out on simple error back propagation training algorithm which is, in turn, developed from gradient-descent optimization technique. The simplicity of this algorithm has led to its appeal in the ranks of civil engineers and software applications. Training of this algorithm is relatively straightforward as it involves adding up of weights of interconnected links and nodes. With each iteration, training riles specify how weights are modified.

Artificial Neural Networks (ANN) is an approach to functioning of human brain in terms of information processing. Knowledge Based Neural Network (KBAAN) which is used for empirical learning is based on ANNs. Among all empirical learning systems available across a range of domains, ANNs have proved superior in almost all respects. ANNs are superior when it comes to meeting data-intensive challenges and where rules for solving problems are either unknown or cannot be easily expressed. ANNs are generally considered robust and relatively immune to errors that can creep in with data processed. ANNs can be applied to a range of problem categories, viz., clustering, function approximation, prediction, pattern classification, retrieval by content and through process control, etc.

Figure 2. Artificial Neural Networks (Feed Forward)

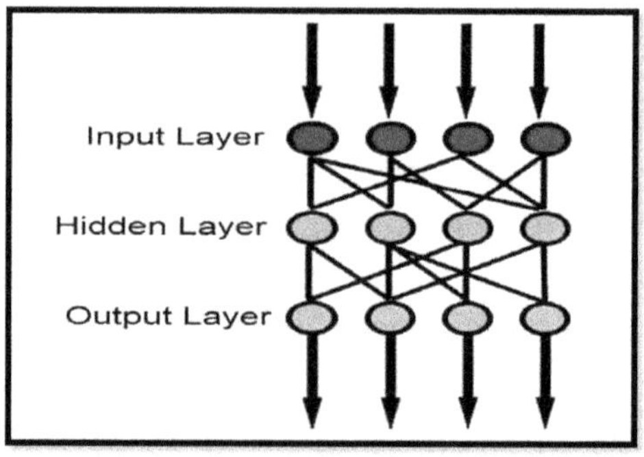

In construction sector, ANNs have found applications in areas concerning cost of construction, budgets, and tender bids. AI has been used to predict cash flow in construction projects, maintenance schedules, and workforce productivity. Researchers have developed ANN to help select cranes. The model that was developed was tested for estimating costs of RCC structures and also for planning work breakdown schedules of projects.

Fuzzy Systems

Fuzzy Logic systems are designed to function with either imprecise or incomplete data. Applying conventional logic and mathematical set theory, an object can either be a member of a particular set or otherwise. Fuzzy Logic allows considering a value anywhere within the range of 0 to 1. Fuzzy models are, therefore, able to work on statements that are not well-defined. In natural language systems, fuzzy logic is often applied to solve a broad range of problems. While Fuzzy systems are not designed to inherently offer capability to learn and memorize, this disadvantage is offset by coupling it to other models to produce hybrid versions. For example, Neuro-fuzzy systems are developed by combining Fuzzy systems and neural networks.

In construction sector, Fuzzy systems can be used to analyse and collate knowledge and information on engineering problems and solutions. The information analysed can then be further developed and used in decision-making processes. Fuzzy Logic enables mapping and modelling of imprecise data in a manner that can enhance quality in real-time applications.

Figure 3. Fuzzy Model Main Components

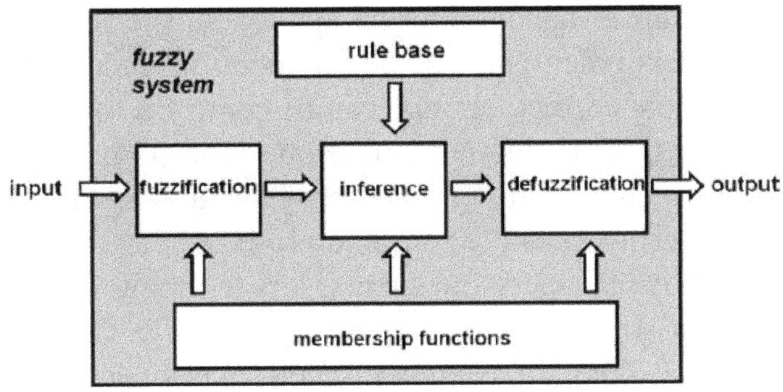

Figure 3 is a classic illustration of how Fuzzy Logic systems work.

ROOM FOR AI AND MACHINE LEARNING APPLICATIONS IN CONSTRUCTION SECTOR

AI and Machine Learning can contribute to construction industry in many ways. To detail or even discuss the ways in which these can be used in construction industry would be well beyond the scope of this book. A simple illustration of AI in action is how mail filters operate these days. Even a few years ago, mail boxes would be inundated with useless emails or even emails with malicious content. Today, most email service providers have spam filters powered by AI that are very good at keeping spam emails from reaching inbox.

Construction projects have myriad issues, changes, processes that need to be addressed on an almost constant basis every day. AI can help identify and highlight the top issues that need human attention and intervention. Critical issues and safety concerns that may otherwise be swept aside by an avalanche of mounting data and information, can be brought to immediate attention by AI that is able to sift through voluminous chunks of data in very short time spans. AI applications can not only help sorting issues and problems but can also assist in operating drones, monitoring and managing machinery, or even help architects and engineers come up with better designs.

The ways in which AI is helping construction companies achieve results are discussed below:

- AI powered scheduling engine can assist construction companies maintain project goals. Managing manpower and material resources can be scheduled for delivery and notifications raised for immediate attention of human handlers by AI applications. Even certain critical handling of equipment such as deployment of cranes or even skilled manpower at the correct juncture can be controlled by AI.
- AI can be used to fly autonomous drones and roving robots that can track and map construction project progress in real-time. Data reaped from such overflights and on-ground close monitoring by robots, can then be used by managers and supervisors to monitor productivity and initiate action whenever and wherever necessary. The possibilities to synchronize design parameters with metrics from actual data revealed from surveys of project progress, are numerous
- AI applications can help companies to create a safe and secure work environment for personnel working at locations where there accidents may happen any time. Applications such as visual monitoring and AI combined with Autodesk BIM 360 can help safety managers to address safety issues at work. Studies have shown that an overwhelming majority of construction-related accidents have been caused by events that, if identified and addressed in time, could have been avoided. Researchers argue that AI can help address these concerns effectively.

THE RELATIONSHIP BETWEEN THE CONSTRUCTION INDUSTRY AND AI

The relationship between construction sector and Artificial Intelligence is influenced by five factors. These are detailed as follows:

Saving Time on Surveying

AI can help fly drone autonomously over spaces and areas that may be inaccessible or danger for humans to physically venture. AI can not only accurately map remote locations, but can do the work in short span of time than if human labour been involved.

Artificial Intelligence and Building Information Modelling

Building Information Modelling has revolutionized the way construction companies approach project works. Building Information Modelling allows sharing of knowledge throughout a project lifecycle and even beyond. Coupled with AI, Building Information Modelling can go a step further in providing detailed insights on interrelated information and data collected over a period of time on activities related to construction works.

Planning Construction

AI can be used in the initial stages of a construction project. Using AI, autonomous equipment can survey a location to create detailed maps for design and producing plans for construction. Usually, such activities take up several days to even weeks. With AI, such activities can be covered in a space of only a few days or even less. As a result, construction companies can save both time and money, and yet achieve an acceptable level of accuracy that can be further fine-tuned through minimal human intervention.

Managing the Project

AI can actually help manage project works in real-time. In managing HR related issues, AI can help allot tasks depending on availability of personnel and attendance details.

Embedded Sensor With Structures

AI can work on data harvested from sensors embedded within walls of buildings, home appliances, and furniture or even in décor. Not only is AI able to predict and sense occupancy based on proximity data, it can also adjust heating and lighting, and perform any other action that would have otherwise required a human to perform. In case of construction projects, embedded sensors can report back a wide range of data that can include data on temperature, wind pressure, humidity, stress and strain, etc. Depending on certain conditions, AI can trigger appropriate warnings or alerts for human intervention to take place in order to address any issues before such can develop into problems require a far greater amount of human resources and costs.

AI in Architectural Design

Although human mind still holds an advantage over machines when it comes to designing and being generally creative, AI can sift through many billions of data tirelessly and come up with radical solutions. Architectural designs that would have otherwise required extensive hours to test and fine-tune, can easily be handled by AI. While designing a particular structure, AI can also come up with detailed calculations related to material and human resource costs, timelines involved, logistics, etc.

Safety

Construction industry is sensitive to safety issues and concerns. Globally, construction industry fatalities account for a significant portion of losses incurred in terms of valuable human resources lost. AI is now being used by many construction companies to detect and track human resources in and around hazardous areas at construction sites. Microsoft, recently, unveiled a new AI platform that enables construction companies to operate sites safely and productively.

APPLICATION OF ARTIFICIAL INTELLIGENCE IN CIVIL ENGINEERING

As discussed previously, the role of AI in civil and construction engineering has a significant dimension. Project evaluation, diagnosis, prediction, decision-making and optimization are some of the important areas where AI has a decidedly lasting impact.

Role of Smart Materials

Structural Health Monitoring

With embedded sensors in vital infrastructure installations such as bridges, AI can analyse data collected from the sensors and create alerts for maintenance crew to carry out preventive maintenance. For structural designers, engineers and architects, AI can be an invaluable tool to create better designs that can withstand geological and meteorological influences.

Self-Repair Materials

In many industrial applications, self-repair materials are used. For example, in structures underwater or in space, any break in material of a structure can be repaired through uncured resin which flows to seal open spaces from embedded tubes within structures. With AI, engineers can direct flow of resin, in advance, to areas that are predicted to fail at different points of time.

Structural Engineering

As explained in previous section of this chapter, AI has an important role to play in structural engineering. Collating and analysing data followed by AI-predicted alerts can go a long way in reducing the possibility of costly maintenance and repairs. In structural engineering, the role of AI is not only to predict areas that would repairs and replacements well in advance, but also to assist in optimizing time and resources necessary for such to be completed in a manner that is aligned to intended design parameters.

Waste Management

The subject of waste management has been gaining momentum in the wake of worldwide calls for a better environment. Arranging to dispose waste is not only a complex task, but is also expensive and requires time. While the use of smart materials with embedded sensors and micro-processors, AI can be used to automate waste reduction and disposal processes.

Concrete Mix Design

Mixing concrete is a design based on several important parameters such as fresh concrete, desired strength of the finish product and durability. These factors are governed by the ratio of cement and water. AI can help control the ratio by analysing the coarseness of mortar and the desired parameters involved to produce finished hardened concrete.

Estimation

Artificial Neural Networks have been found to excel in situations where decision-making processes are influenced by analogous examples of data available.

Role of Artificial Neural Networks (ANN)

Neuromodex – Neural Network System for Modular Construction Decision Making

This model assists in adopting an approach on whether to involve conventional “stick-built” method or to rely on modularization to a certain extent in a given project. Such decisions are influenced by several factors involving project location, human resources, project characteristics, and project risks. The neural network is trained using data collated from a wide range of engineering and construction projects.

Initial Design Process Using Artificial Neural Networks

The preliminary stages of designing a model are vital. Artificial Neural Networks can help predict an optimized design based on tensile requirements, costs and resources required from a input parameters ranging from dead load, live load, span, grade of concrete and steel required.

ANN in Non-Destructive Testing

In non-destructive testing processes, two methods are popular among architects and engineers. The Schmidt Test Hammer Method and Ultrasonic Pulse Velocity Method. Researchers have observed that for correlating Non-Destructive Testing (NDT) parameters to strength of a structure, ANN can be an appropriate choice. This is because ANNs do not require predefined mathematical relationships to exist beforehand.

ANN in Predicting Large Deflection Response

Researchers and engineers have recently been conducting studies to use ANN in evaluating responses of large deflections in fixed or immovable rectangular plates that are subject to patch loading. Sigmoidal function in the range 0 to 1 has been used with error back propagation algorithm in order to assess relationship between patch size, pressure coefficient, aspect ratio of input plate alongside central deflection, bending and membrane stresses at several points of a rectangular plate.

Planning Construction Projects

AI applications can be used to plan various stages of construction, production and fabrication processes including taking into consideration preconditions and various effects. With AI, new sets of actions can be inserted into project plans with a view to optimizing resources depending on any changes in environment, logistics, human resource allocation and resources alterations.

Construction Robot Fleet Management System Prototype

Given hazardous conditions in which construction workers are often required to work, robots can be a safe alternative. Powered by AI, robots can be used to undertake a variety of dangerous or repetitive actions that would have otherwise necessitated use of workforce in conditions that can severely undermine safety of humans.

Bridge Planning Using GIS and Expert System Approach

Construction processes involving roads and bridges in often difficult terrain and under adverse environmental conditions, require GIS to collect data. With the help of AI, huge volumes of data can be sifted, sorted and analysed for creating better designs. Simulations using digital models can be carried by AI, thereby, enabling architects and engineers to conserve resources and time.

Artificial Neural Network Approach for Pavement Maintenance

Pavement maintenance crews can gain a significant advantage with the help of AI. Using existing data and past records, AI can predict an optimized course of action to conduct pavement repairs.

ANN for Environment Health and Safety (EHS)

With automation and robots, AI can make hazardous workplaces that much safer for humans. Humans are prone to errors. However, robots are not. Autonomous machines controlled by AI can not only accumulate and analyse vast quantities of data, these can do so at a fraction of the time it would require humans to do so. That apart, AI-controlled robots can perform any set task repetitively and without tiring out as a result of fatigue as in the case of human beings.

Tidal Forecasting

With construction of ports, ferries, and jetties gathering pace around the world, construction engineers and architects are aware of tidal waves and water movements that can affect such projects. Studies have shown that Kalman Filtering Method when used in conjunction with AI can help engineers and planners determine tide levels to a significant degree of accuracy.

Earthquake-Induced Liquefaction

Earthquakes are one of nature's fearful destructive forces. For centuries, human civilization has experienced the deadly effects of earthquakes on cities and magnificent structures built by human hand. Earthquakes cause destruction through either structure failures or foundation failures caused by liquefaction.

AI can help design better structures that can not only withstand earthquakes, but can also mitigate the debilitating forces on foundation of civil structures.

REINVENTING THE ROLE OF PROJECT MANAGER IN ARTIFICIAL INTELLIGENCE ERA

Scientists argue that the human brain still retains an edge over AI. However, with increasing technological advancements, AI is slowly but steadily gaining traction in areas where humans were considered indispensable. The importance of the role of project manager in construction projects cannot be over emphasized. That said, however, project managers are often engaged in repetitive tasks that use up considerable resources and time. AI can simplify and rationalize work for project managers leaving them to contemplate on issues that require lateral thinking and unconventional approach to yield competitive advantage.

In the following sections, the areas in which AI can be useful for project managers in construction industry are brought to attention for readers.

Integration Management

Intelligent bots, or AI-assisted robots, can respond to voice requests. With the help of such devices that can also function autonomously, project managers can instantly obtain and view information through speech activated systems rather than through tactile interface systems. AI can set up alerts based on historical weather patterns that can affect project work schedule. It can also help predict optimize resource usage based on a range of factors that may change without warning.

Scope Management

Work Breakdown Structure (WBS) can be optimized and made more comprehensive with the help of AI. Existing organizational knowledge can be used by AI to offer project schedules and key performance indicators all based on historical data analysis.

Cost Management

Cost estimation is a vital exercise that determines success or failure of any project. Since, estimation of costs depends on understanding design and engineering requirements involved, the role of AI can be augmented through analyses and accumulation of knowledge on available data on assorted cost models.

Time Management

As discussed in previous sections, the value of AI in managing time in construction projects is inestimable. Using ANNs, project activities can be sequenced based on requirements at various stages. AI can facilitate in readjustment of project schedule based on dynamic changes influenced by assorted factors ranging from allocation of human resources through to logistics and even weather and climatic patterns.

Risk Management

Using a popular simulation technique known as the Monte Carlo simulation, AI and ML combined can help project manager assess risks and carry out simulated tests to quantitively determine the extent of risk that a particular project may be exposed to. Fuzzy logic is often used to determine probability of risks in infrastructure projects. AI, in such cases, is instrumental in sending out alerts to project managers on available upstream opportunities or positive risk, and downstream obstacles or negative risks in real-time.

Quality

Using cognitive technologies, it is possible for project managers to assess quality of work and obtain insights to focus on ways to optimize costs and boost productivity. However, cognitive technologies are not limited to the field of actual project implementation alone. With the help of natural language processing, AI-assisted bots can help project manager get a quick summary on important terms and applicable conditions in contract documents. This ability of AI to scan through contract documents and ferret information vital for legal as well as contractual obligations, can help save valuable time and resources. As a result, quality of work can be enhanced through maintaining a clear channel of communications through to clients of construction company, while at the same time freeing project managers to allocate more time to production and construction works.

Stakeholder

Sentiment analysis algorithm is a recent technological advancement that has revolutionized the way machines is able to understand and predict sentiments of humans. AI that uses sentiment analysis algorithms can help project managers analyse text of emails from customers to quickly grasp the core issues and determine ways to engage in meaningful and productive communications with all stakeholders essential to a project.

Human Resources

Human resource is perhaps the all-important factor in any business or industry. AI-assisted human resourcing bots can help project managers identify and locate essential personnel and workforce. AI-assisted bots can even go through online job portals and online professional communities such as LinkedIn, to reach out to potential candidates. AI can handle billing, assignments and accounts related to human resource allocation and even work entrusted to freelancers and independent consultants and contractors.

Procurement

Procurement of resources and materials from reliable suppliers and ensuring timely delivery is a vital activity necessary for keeping any construction project on track. AI-assisted bots can not only conduct risk analysis and track performance of suppliers and recommend appropriate ones to planners and project managers, such AI-assisted bots can also initiate conversation with suppliers about orders through convenient web application interface. AI can also determine deviations and compliance issues for supplies made to ensure that a high standard of quality is always maintained.

REFERENCES

Adeli, H., & Hung, S. L. (1995). *Machine Learning—Neural Networks, Genetic Algorithms, and Fuzzy Systems*. John Wiley &Sons, Inc.

Russell, S., & Norvig, P. (1995). *Artificial Intelligence: A Modern Approach*. Prentice Hall.

Schalkoff, R. J. (1990). *Artificial Intelligence: An Engineering Approach*. McGraw-Hill.

Chapter 10
Recognizing the Significance of Human Resource Through Human Resource Information Systems in the Construction Sector

ABSTRACT

The value of human resources in every commercial and industrial sector cannot be over-emphasized. Even in the age of automation, management of human resources has never been more significant or relevant. Information technology can help in management of human resources such that companies can make optimum utilization of skills and experience while keeping costs low. In this chapter, the authors touch upon the various aspects of human resource information systems and the usefulness of such in managing human relations across all levels of employment.

At the height of industrial revolution in early twentieth century, Human Resource came to be regarded as one of the most serious issues that were to go on to shape human civilization in developed countries. Human Resource Information Systems (HRIS) consist of specific computer applications together with database systems that are designed and installed to support personnel and human resource management functions. Human Resource Management

DOI: 10.4018/978-1-7998-5291-9.ch010

Professionals (HRMP) use such systems not only for compliance and regulatory practices, but also to bolster productivity, management payroll and employee benefits, and reduce costs associated with acquiring additional human resources. HRISs usually offer a diverse report-generation system and built-in analytical tools that help in managing human resources effectively. The value of service rendered by HRISs to organizations cannot be over-emphasized. HRISs help in supporting strategic business goals such as identifying and helping in recruiting personnel with right set of skills required for a range of job functions. For construction companies, the potential of HRIS is significant as project managers and planners can quickly provision human resources based on dynamic project requirements.

The fundamental function of HRIS is to manage relationships between an organization and its employees. With HRIS, organizations can not only post advertisement to highlight vacant positions, but can also scan, sort and store resumes received. HRIS enables companies to carry out performance appraisals, compute and manage compensations and employee benefits, as well as maintain a proper record listing employee skills, competencies and experience. For employees, HRIS serves as a portal for seeking new career growth paths within the organization and to explore training programmes for skill augmentation. HRIS enables organizations to seek and identify skilled employees for specific tasks. It also allows generation of reports and analytical data assessing information on linguistic skills, leave information and other relevant data. Enterprise Resource Planning software has wide use in Human Resource applications. Organizations can manage and monitor individual employees at various stages of their professional engagements within enterprise. Since, corresponding changes can be viewed as tranches of transaction information, HR Management information is well-suited for adaptation in ERP infrastructure. HRIS can be programmed to include consistent parameters related to labour and knowledge economy. This helps HR managers to employ workers for project activities. On account of such reasons, HRIS (Kovach, 1999) is considered distinct from ERP systems and are often used as separate product with specific options such as customizable statistics on employment unique in each industry or business segment.

Software systems that are useful for Human Resource applications are marketed under various generic names. These include *e-Human Resource Management*, *Human Resource Information Systems*, and *Human Resource Management Systems*. The differences between these three classes of HR systems although small, are yet significant. Transitioning from automation of HR services, also otherwise termed as *transactional systems*, to extending IT support for HR Management Information Systems, is how HRIS is described. In addition to this, HRIS is described as a database or a network of linked databases supporting a range of HR functions and data.

DEFINING HRIS

Human Resource Information Systems (HRIS) use Information Technology to manage human resources. HRISs are integrated systems that are used for gathering and analysing information on human resources deployed at various levels in an organization.

WHY IS HRIS REQUIRED

In modern workplaces, HRIS enables HR professionals to automate routine HR activities. This allows HR professionals to focus on key strategic goals such as talent management and develop leadership programmes to improve human resources. There are several other incentives for implementing HRIS (Ngai, 2006). These are as follows:

- Develop programmes and frame policies for human resources.
- Manage and improve functions such as nominations, promotions, transfers, statutory employee benefits, leave compensation, etc.
- Provide data and submit returns to government and statutory bodies.
- Analysing, extracting and storing relevant data in usable chunks of information to improve decision-making.
- Generating array of relevant and detailed reports accurately and in real-time.
- Significantly improve employee satisfaction by engineering better responses quickly and efficiently.
- Scale up competitiveness through improved HR functions and processes.
- Provide reports at global and granular level to provide comprehensive information at any stage.

For organizations seeking to leverage their existing human resources, HRIS (Hussain, 2007) and Information Communication Technology are worthwhile investments that can prove to be beneficial for employers as well as employees.

Human Resource Management in Construction of Sustainable Project Development

Human resource is vital for success in any business venture, project or production. Lack of skilled human resources at appropriate junctures can lead to operational loss and can have a negative impact at organizational level (Carter, 2000). For this reason alone, HRIS is considered a significant investment in the modern era where organizations need to survive in intensive competitive environments.

Project Success

The factors that affect success of a HR Management are as follows:

- Clearly defined project objectives.
- Detailed breakup of project schedule.
- Clear communication between all stakeholders.

Identifying and negotiating with manufacturers of hardware and computer software to make necessary arrangements so that appropriate tools and technologies are at hand at every stage of project work.

- Clear and unambiguous acceptance by client or end-user at completion of every earmarked project milestone.
- Timely monitoring and feedback.
- Recruitment, selection and training of human resources.

Policies and Practice

HR Management policies need to be aligned with an organization's overall objectives. When HR policies and practices are integrated in broader framework of an organization's strategic aim, it is known as vertical strategic integration. Horizontal integration occurs when policies and practices are aligned with each other. In certain cases, organizations need to focus managing by projects instead of focusing primarily on overall organizational objectives. While this may appear to be counter-productive, intense competition in markets demand that organizations adopt a more flexible posture and be more receptive to requirements of employees.

Training

The usefulness of training cannot be overstated as without it overall quality of work shall automatically go down. Managers, senior executives, engineers and supervisors must be encouraged to share and transfer their knowledge and skills with their subordinates. In construction industry, a strong emphasis on training must be placed as this sector relies on skills of its human resources. It is in this area that HRIS can play an important role in measuring, monitoring and augmenting quality of output generated by human resources of a construction company or, for that matter, any company. For workers engaged in construction sector, training can be imparted while on the job or off it, as in off-the-job training.

Skilled and Unskilled Workforce

Construction sector employs human resources to comparably larger extent than any other industry. A construction project involves people with a wide variety of skills and interests. A typical construction project involves clients, architects, designers, engineers, managers, administrators, consultants, supervisors, foremen, contractors, skilled and unskilled workers. Although construction sector is one of the largest employers in terms of human resources with comparably low reliance on technology, HR management is an aspect that is not often given sufficient attention.

Challenges Facing Human Resource and Management

There are several issues that construction companies are aware of that affect the functioning and overall operational efficacy of an enterprise engaged in civil construction works and projects. These issues are as follows:

1. Replenish or replace ageing but skilled workforce.
2. Recruit adequate number of skilled human resources to meet rising demands.

3. Improve image of construction sector to make it a sustainable career option for the youth.
4. Monitor and manage supply of skilled and unskilled workers in fluctuating economic conditions and volatile industry situations.
5. Training for supervisors and on construction management should be made available and easily accessible.
6. Apprenticeship Training Programmes should be made more available and accessible.
7. Improve basic skills among workforce to meet rising demands.
8. Train workers on newer technology for improved productivity in both short- and long-runs.
9. Be sensitive and extremely responsive to stringent health and safety requirements not only to meet government regulations but to protect human resources and, thus, safeguard an enterprise's own business interests.

Broadly speaking, construction companies face four types of challenges. These challenges remain as pressing today as these would be in future, too. Thus, a focussed and an objective approach needs to be formulated by individual construction companies keeping mind their own organizational interests, to overcome the hurdles posed by the following:

- To seek a balance between demand and supply of skilled workforce.
- Create and maintain a sustainable and dynamic apprenticeship programme.
- Continuous development of a flexible training programme to upgrade skills of human resources of organization.
- Enabling quick response strategies to mobilize workforce to locations where relevant skillsets are required.

Employee Resourcing Challenges in Construction Sector

Resourcing of employees is a holistic activity comprising selection, recruitment and deployment of staff as per requirements of different operating sections in an enterprise. Companies operating in construction sector face stiff challenges in recruiting skilled workers and trained personnel. Considering dynamic project nature, resourcing of skilled workers and trained personnel require close attention and awareness of market dynamics and competition posed by competing organizations. A flexible approach is, therefore, required of HR Departments, managers and supervisors in identifying, recruiting, selecting and training of human resources either from outside organization or internally. The advantage that HRIS offers HR managers is to be able to identify, recognize and assess individual performance and training requirements. Considering that high employee attrition can adversely affect business operations, organization with HRIS can meet the challenge of gain long-term benefits through development and retention of personnel, head on.

The Benefits of Human Resource Information Systems

Human Resource Information System is a term that describes software and related systems designed and developed for supporting human resource management functions in large enterprise and organizations. The authors, Broderick and Boudreau, once described HRIS being a composite of databases, computer hardware and software that is used for aggregating, recording, storing, managing and presentation of human resource data in a form that allows stakeholders to formulate decisions and base strategic goals based on reports produced.

Broadly these researchers have identified three types of software applications that can meet human resource management requirements:

1. Reporting / Tracking / Transaction Processing applications that are suitable for a high volume of decisions.
2. Expert Systems that help improve decision-making processes using rules formulated using analyses and expert decisions made over time.
3. Decision-support Systems which aid in improving decisions requiring change in rules that are otherwise not defined nor results known or predictable.

Modern HRIS offer advanced features that enable HR managers to store and analyse endless volumes of data relating to human resource skills and cumulative experience in an organization. Such HRISs are also able to offer complex and sophisticated reports that map employee skills and preferences to vacancies in any upcoming positions or projects. Aspirations of employees can also be recorded and mapped by such HRISs to provide holistic and comprehensive profile of an employee and to gauge suitability for recruitment in projects on a case to case basis.

Most modern and latest versions of HRISs offers employees to update their personnel information, provide their own timesheets, examine benefits available to them, apply for holidays and even enrol on company-listed training programmes. Such integration of important HRM activities helps promote an environment where recruitment and retention of personnel is made possible through quantification of total compensation packages and figures along with automated recruitment facilities. All these features that a HRIS offers are aimed at freeing HR professionals to focus more on strategic goals and to align human resources that are available to them into a potential source to maximize profitability of organizations. The key advantages of HRIS are as follows:

Figure 1. Overall benefits of HRIS

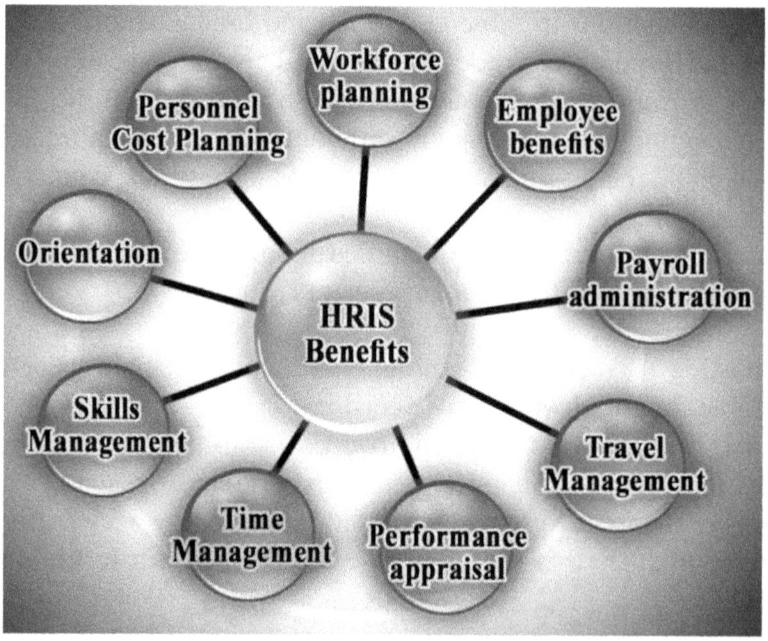

1. Improve human resource management for sustaining in increasingly competitive domestic and global markets.
2. Include strategic human resource information along with operational / transactional human resource. information to address emerging wider issues affecting profitability of organization.
3. Make employees active players in HRIS
4. Redefine and reengineer human resource department into a strategic wing of operations.

The advantages and benefits of HRIS are listed as follows:
Advantages for Management:

- Positive effect on decision-making processes.
- Reducing costs and a better control on budgets.
- Clarity in business goals and strategies.
- Optimized recruitment and retrenchment of staff and personnel.

Benefits for HR Department:

- Unified single database system for all personnel and staff.
- Digitization of records leading to more accurate tracking and minimizing of errors.
- Real-time updates to database.
- Minimizing chances of redundant information.
- Aligning HR goals and operations with government regulations and laws.
- Enabling set standards for business processes.

Benefits for Staff and Personnel:

- Better time management.
- Access to data independently.
- Access to individual personnel records any time of day or night.
- Improve morale of staff and personnel.
- Automatic alerts to business and organizational events.
- Attend internal training programmes via Internet.

Importance of HRIS in Increasing Competitive Edge

Organizations, worldwide, are aware of the deep impact that human resources have on their profitability. In domestic and global markets, companies that can retain their competitive edge are able to withstand adverse market forces and survive. Human resource is a vital ingredient essential to success or failure, if ignored, for any organization. For companies working in construction sector, human resource ranks as one of the top important factors affecting profitability and competitive edge. A well-designed HRIS can be an important management tool for setting and maintaining HR policies and practices, and alert HR managers to any gaps and issues that need immediate attention.

Figure 2. Road map business competitiveness

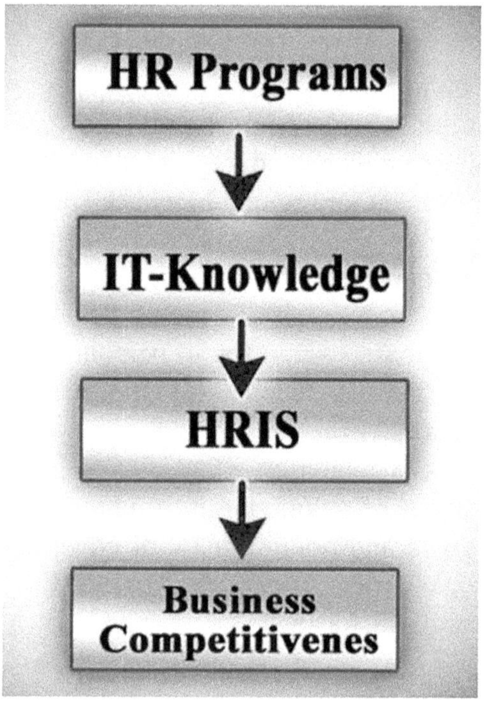

HRIS Model

Organizations use HRIS as a tool to address and solve various issues and problems associated with human resource and staffing. In many cases, organizations have been known to shift from process-centric human resource management to individual-centric human resource management. This paradigm shift in the outlook of HR policies and practices have enabled companies to adopt an integrated approach towards staff and personnel.An example of HRIS model is detailed in the figure shown as follows:

Figure 3. Model of Human Resource Information System (HRIS)

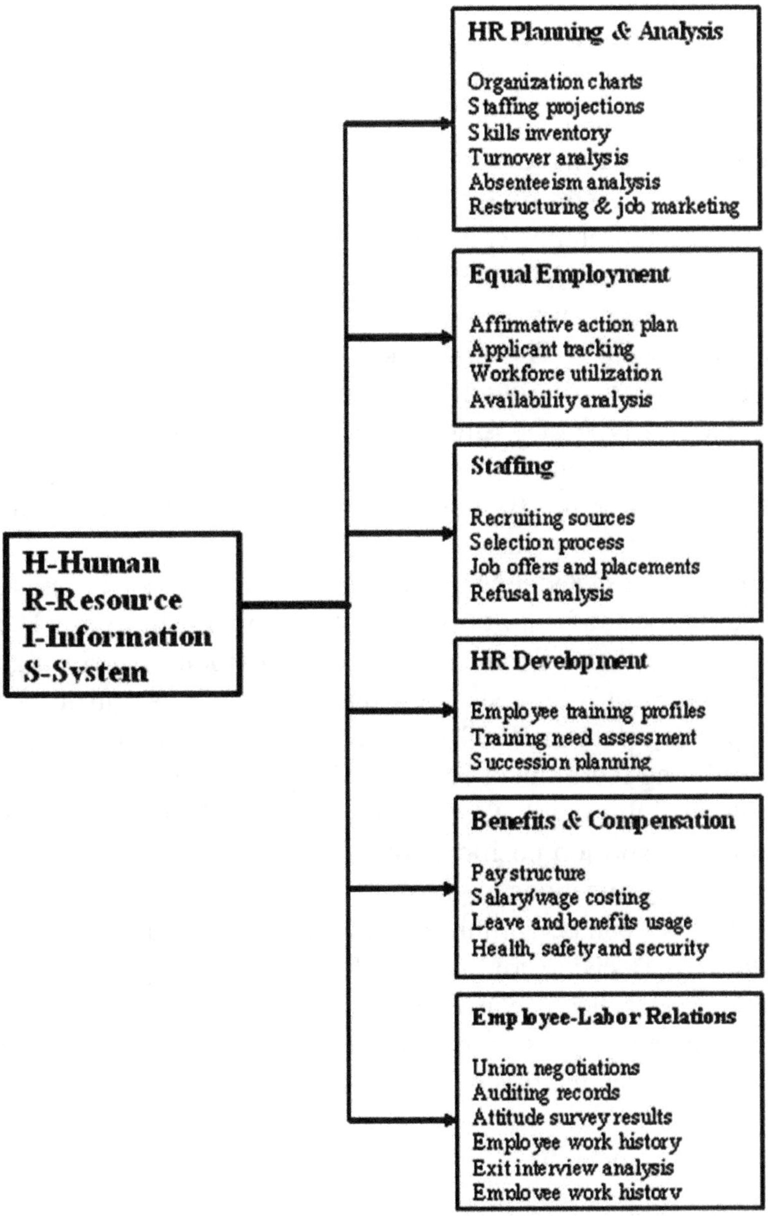

Functions of HRIS

HRIS is an integrated and responsive solution that supports all essential and key HR functions. These are categorized as follows:

1. **Creation and maintenance of individual employee records:** Employee data records are created when HR administrator enters information in HRIS. Throughout employment, individual records is kept up to date and maintained in the system.
2. **Ensure legal compliance:** Records and data in HRIS can help organizations comply with government regulations and laws. In matters of law, compliance in timely and accurate manner can help in avoiding costly legal expenses and penalties.
3. **Human Resource Planning and Forecasting:** HR managers can base longer term human resource plans from data available in HRIS to ascertain number and type of positions that are open, skills and competencies required, retirement eligibility and employee benefits. Together these data can be used by HR managers to develop recruitment, training and skill development plans.
4. **Talent and Knowledge Management:** HRIS can be used by organizations to seek and identify appropriate employees within the ranks. Employees can be identified and trained for specific positions that are aligned with an organization's immediate or long-term goals. Acquisition of talent by applying data in HRIS knowledge banks, can significantly aid in streamlining recruitment costs on a long-term basis.
5. **Strategic Alignment:** With human resource data mapped to skillsets, experience and job positions available or to be filled in future, HRIS can help organizations align strategic goals in a manner such that HR related costs can be optimized without compromising productivity.
6. **Enhanced Decision-making:** HRIS enables managers and planners to make quick decisions. Since information is digitized and can be accessed instantaneously, decision-making processes can be quickened and made relatively error-free based on data available.

Components of HRIS

Several components in HRIS can exist depending on design and development made by software developers. The more common among these components are discussed as follows:

1. **Database -** At the core of every HRIS is a database. A database is a central repository system that is used for storing all kinds of information.

Without a database, a HRIS would not exist and it would be impossible to store, segregate, collate, and quickly produce records and information from records and tables.

2. **Labour and Time Management -** Time and labour management are not only time-consuming; these also require complex calculations involving participation from all levels of management at times. HRIS allows employees to feed their own attendance information for the data to be verified by HR managers and onward payroll processing or requests for holidays.
3. **Payroll Function -** HRIS enables organizations to streamline the process of issuing pay checks, calculate and make statutory deductions and comply with laws of any particular state or region while conforming to more than one level of taxes.
4. **Benefits -** HRIS enables employers to maintain health benefits and retirement investments through a single software. This allows organizations to focus on more important activities and business processes, while at the same time paying complete and undivided attention to interests of staff and personnel.
5. **Employee Interface -** Some, though not all, HRIS allows employees to have limited user access. Such limited user access may involve updating of personal information, examining pay scale, alter retirement benefit program enrolments, etc.
6. **Recruitment and Retention -** HRIS performs a crucial role in recruitment and retention of skilled staff and personnel. With HRIS, organizations can not only recruit as per work requirements, staff and personnel can also be trained and upskilled to fill in positions in projects requiring specific skill sets.

Types of HRIS

HRIS can be of four types. These are detailed as follows:

1. **Operational HRIS -** This type of HRIS enables HR managers to work with information on employees as well as align HR practices with government regulations. This type can be further sub-divided into two types:
   a. **Employee Information Systems:** Employee Information Systems enable organization to keep track of all personal and professional

employee data. Work experience and work history related information are also stored.

b. **Position Control Systems:** Position Control Systems are of use to HR managers who require information about any unfulfilled position within an organization. HR managers can then work their way to identifying the requirements of that position, job title involved and information on any existing employee assigned to that job role.

2. **Tactical HRIS** - Tactical HRIS enables HR managers in decision-making processes that require decisions to be made based on recruitment, job analysis, training, development, benefits and compensation. This type of HRIS can be further sub-divided into several other types which are as follows:
   a. Job Analysis and Design Information System
   b. Recruiting Information System
   c. Compensation and Benefits Information System
   d. Employee Training and Development System
3. **Strategic HRIS** - Strategic HRIS offers features that enables organizations to focus on workforce planning and labour negotiations. This type of system can be further sub-divided as follows:
   a. Information Systems Supporting Workforce Planning
   b. Specialized Human Resource Information Systems Software
4. **Comprehensive HRIS** - Comprehensive HRIS is a type that has evolved with diverse data on human resources in a single database or coordinated centralized database systems. Such type of HRIS allows application software to produce a wide variety of reports from data available in Comprehensive HRIS.

HRIS Process

The processes which HRISs engage are as follows:

- Recruitment and selection
- Training and development
- Performance and evaluation
- Promotions
- Maintaining records of personal information
- Compensation and benefits
- Prospects and career path

Figure 4. HRIS Process

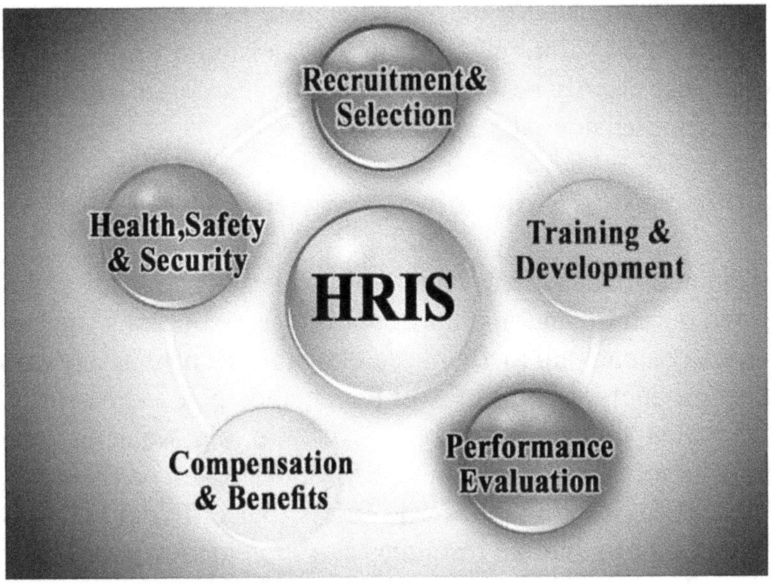

Structure of HRIS

The various features and functional attributes of HRISs have been discussed in previous sections of this chapter. Individual HRIS may differ in several ways depending upon business sector involved and software developer marking HRIS products. Companies such as SAP bundle HRM as a component of ERP. For example, SAP HRMS is one of the bigger modules of SAP R/3 system. Yet another group of companies can market HRM products covering a single set of features. Regardless of specific sets of features with which HRISs may be marketed, distributed and sold by software manufacturers, following are generic or commonly used modules that are found present:

- Collection and monitoring of employee information.
- Maintenance of records.
- Payroll module.
- Benefits and administration.
- Training module.
- Performance tracking.
- Employee self-service and management.

Given the wide variety of HRISs available in the market for sale, it is important that the correct type be purchased to maximize benefits of a computerized HRM solution. As such, the decision to select and purchase HRIS should necessarily be a strategic one with careful thought given to the end-use and long-term goals.

Introduction and Replacement of HRIS

Human resource management is a complex activity. Not all organizations can afford to purchase and maintain sophisticated solutions such as HRISs. There are many small and medium scale organizations that rely on standard spreadsheet products, e.g., Microsoft Excel, and open source software such as Orange HRM to meet their objectives. Certain websites, e.g., www.comparehris.com, offer advice on HRIS and HRMS products. It should, however, be borne in mind that HRIS offers four different features that are integral for managing human resources at a strategic level for organizations big or small. These are as follows:

- Data Management
- Human Resource Management
- Talent Management
- Organization Management

The above factors are crucial to compute returns on investment or ROI coefficient. HRIS being costly investments, computing ROI factor for a strategic decision in choosing the appropriate HRIS product is important and a worthwhile investment.

Importance of HRIS

The importance of HRIS cannot be overstated. For construction companies with diverse portfolio in terms of construction projects and at different geographical locations with own local laws on labour and staff welfare, it is critical that a proper HRIS product be evaluated and deployed as quickly as possible. For the reasons discussed in the previous sections of this chapter, HRIS can help organizations uncover valuable human resources that would have otherwise been not found in time or not at all. The value of human resource remains a key significant aspect in any commercial or industrial venture. Even more so in the case of construction sector. Well-trained and well-motivated staff and welfare can mean profitability even in times of economic hardships and slowdown in the market. For this reason alone, the contribution and effect of Information Communication Technology in construction sector remain invaluable and shall continue to be so even in future.

REFERENCES

Carter, M. (2000). Control shift. *People Management*, *6*(23), 26–30.

Hussain, J. W. (2007). *The use and impact of human resource information systems on human resource management professionals*. Academic Press.

About the Authors

Bithal Das Mundhra, as the patriarch of one of India's leading construction companies, has been at the vanguard of the Indian civil construction sector. His contributions in the industry have seen him honored with the prestigious membership of "Fellow of the Royal Institution of Chartered Surveyors" in 2006, and with the award of "International B.I.D. Quality Summit Award, New York", Business Initiative Direction, New York in 2009. Dr. Mundhra is widely known within industry circles as an entrepreneur who has seldom shown reluctance in adopting change when necessary. By encouraging his employees to share their visions and ideas, he has repeatedly shown his strong affinity for team work and perseverance. His efforts have been clearly visible in his steering Simplex Infrastructures Limited through periods of globalization and turbulent waves of recessions that have swamped companies in the Indian civil construction sector. For his efforts aimed at modernizing almost every operational aspect of his company, Simplex Infrastructures Limited, he was recognized as "World Leader Business Person" by World Confederation of Business in 2010. He was made "Fellow of the Indian Concrete Institute" by the Indian Concrete Institute and "Honorary Life Fellow" by the Institution of Engineers (India), Bangalore in 2011. The Global Editorial Team of the Construction Week India Magazine, Mumbai honored him with "Hall of Fame Award" in 2012. In the same year, the Builders Association of India, Kolkata announced their decision to acknowledge his contributions and on his completion of "Fifty years of Glorious Service to Construction Industry". In 2017 at the 13th convocation ceremony of the KIIT University, the Vice President of India, the Hon'ble Mr. Venkaiah Naidu, awarded the D.Litt. Honoris Causa as a mark of acknowledgement for his contributions not only in the field of civil construction & engineering in India, but also for his inputs and focus in Indian culture and literature. In several literary and cultural summits, Dr. Mundhra has often emphasized the need to learn and adopt newer technologies while aided with a common unifying Indian language that is spoken by the many millions of civil construction workers. For the first time ever, Dr. Bithal Das Mundhra has now put pen to paper on how adoption of evolving Information Technology and tools can help businesses in civil construction sector meet operational challenges and survive in increasingly competitive markets. His own company, Simplex Infrastructures Limited, first founded in 1924 by his father the Late Madho Das Mundhra, has survived even in highly competitive globalized markets by adopting modern tools and innovations that many others initially found daunting prospect to consider. Simplex Infrastructures Limited was one of the first among construction companies in India to introduce enterprise resource planning. Developed entirely in-house by a team of dedicated professionals and using latest tools in database, programming, and networking technologies, the enterprise resource planning was a shot in the arm for the company. Over the years, the company has kept pace, too, with the latest in design and drawing tools. Modern versions of computer-aided design software and building information modeling tools are frequently introduced to boost productivity and support interoperability with much of those used by partnering consultants, agencies, suppliers, and clients. It is this ethos that Dr. Mundhra explains in this book in clear and simple terms. The 10 chapters are all aimed at bringing readers up to speed on how Information technology can be used to aid in production and operational management while boosting information and data flow throughout an organization engaged in the business of civil construction engineering.

Rajesh Bose has been in the employ of Simplex Infrastructures Limited for over a decade. As such, his position and proximity with senior members of the management and with those from his team directly managing the company's modern and sophisticated Data Center, has put him in a unique position to look at closely emerging trends in cloud computing, IoT, artificial intelligence and block chain tools in the context of civil construction engineering. While contributing to the chapters in this book, Dr. Bose has drawn on his experience and expertise in working on cloud computing, virtualization, IoT, network, and storage. Over the years of his working closely and under directions of Dr. Bithal Das Mundhra, he has had the opportunity of witnessing modern information technology in action in civil construction sector. While working on the chapters of this book, Dr. Bose has had the opportunity of producing an easy to understand approach on how information technology can help civil engineering and construction firms in staying competitive.

Index

W

www.ingramcontent.com/pod-product-compliance
Lightning Source LLC
LaVergne TN
LVHW061536290926
849162LV00004B/811